Kew Bulletin
Additional Series I

Indigofera (Microcharis) in Tropical Africa

with the related genera
Cyamopsis and Rhynchotropis

By J. B. GILLETT

LONDON
HER MAJESTY'S STATIONERY OFFICE
1958

INDIGOFERA (MICROCHARIS) IN TROPICAL AFRICA WITH THE RELATED GENERA CYAMOPSIS AND RHYNCHOTROPIS

J. B. GILLETT

In the following account an attempt is made to provide a means of identification for all species of *Indigofera* (including *Microcharis*) known to occur in the areas of the Floras of Tropical East Africa and West Tropical Africa and of the projected " Flora Zambesiaca ". Two small related genera, *Cyamopsis* and *Rhynchotropis*, which are easily confused with *Indigofera* are also dealt with. In the course of this work it has been necessary to investigate most, but not all, of the remaining *Indigoferae* of tropical Africa and Arabia, and a number from South Africa, and several new species and varieties from these areas have been included. It has also become apparent that the infrageneric taxa hitherto recognised within *Indigofera* are often too unnatural to be of much value and an attempt has been made to work out a more natural system. It is recognised that this is by no means final, as time has not been sufficient for the large number of South African, Madagascan, Asiatic and American species to be properly studied. In view of the doubtful nature of many identifications in *Indigofera* little reliance has been placed on the literature in working out distributions. Where a species is recorded from a given area, unless an author is specifically quoted, a specimen from that area has been seen in the Kew herbarium, or, if so indicated, from another herbarium. All specimens cited have been seen, except those enclosed in brackets. In the description of new species and varieties information gained from specimens other than the type is enclosed in square brackets (see Kew Bull. **1952**, 367). The abbreviations F.T.A., F.W.T.A., F.C.B., and L.T.A., for certain standard works, and U1–4, K1–7, T1–8 for areas in East Africa, as explained in the preface to the Flora of Tropical East Africa, are used throughout.

The author's thanks are due to the authorities in charge of the herbaria in Brussels, Copenhagen, Dakar, Florence, Lisbon, Nairobi, Paris, Salisbury, Stockholm, Uppsala, Vienna and Zurich for the loan of type and other specimens, to Dr. A. Cronquist for the loan of the manuscript of his treatment of *Indigofera* in the Flore du Congo Belge and to Miss D. Hillcoat for many valuable suggestions.

Certain morphological features of use in classification

(A) Hairs

Two radically different types of hair are found in *Indigofera*. The commonest and characteristic type is *unicellular* and is characteristically *biramous*. It varies from the common *medifixed* form in which the two branches are more or less equal, through *unequally biramous* forms, to the *subsimple* in which one branch is almost suppressed, and the *simple*. More than one of these forms are commonly found on a single plant but no plant has been seen in which all the unicellular hairs are simple. As a general term for this type of hair "*biramous*" which is both accurate and self-explanatory is much to be preferred to "*medifixed*" (which should only be used for those which really are medifixed) and "*malphigian*" or "*malphigiaceous*" which are longer and more obscure. The second type of hair is the multicellular. This is erect, simple and usually gland-tipped. It is found in *Amecarpus* sp. 1, §*Paniculatae* spp. 4–5, §§*Viscosae* all spp. and §§*Atratae* spp. 1–5 and 8γ.

(B) Hyaline scales at the base of anthers

Indigofera has hitherto been described as having anthers without scales at the base. In 1925 I. M. Johnston described a specimen of *Indigofera microcharoides* as *Rhynchotropis curtisiae* largely because he discovered hyaline appendages at the base of its anthers. Actually these appendages, which are flat translucent prolonga-

tions of the connective, are found in all species of subgen. *Microcharis* and in
I. fastigiata in subgen. *Indigastrum*. In some species they occur on all the anthers,
in others only on those of the stamens with shorter filaments, particularly those two
which lie on each side of the median ventral stamen. As these scales are translucent
they are not easily detected when the filament is present behind them, and in a
detached anther they may be mistaken for a broken off portion of the filament.
The best way to observe them is to view the stamen from the side, when the thin
plate projecting from the base of the anther is seen to be quite distinct from the
subcylindrical filament attached to its back.

(C) Spots in the endocarp

A useful diagnostic character is the presence or absence of dark red, brown or
blackish spots in the endocarp. These are often associated with corresponding pits
in the seeds. Dr. C. R. Metcalfe very kindly investigated these spots and found
them each to consist of several swollen coloured cells rich in tannin. It appears
likely that these cells may be prominent, and that the young developing seeds mould
themselves around them, thus becoming pitted.

(D) Simple or unifoliolate leaves

In the classification of *Leguminosae* and other families a good deal of stress has
sometimes been laid on the distinction between "*simple*" leaves, considered not to
be derived by reduction from compound leaves, and "*unifoliolate*" leaves which, by
the presence for instance of a joint and stipel at the top of the petiole, seem to have
been so derived. *Indigofera* usually has compound leaves, but a considerable number
of species, either always or frequently, have unifoliolate or apparently simple leaves.
In some species examination under a hand lens provides clear evidence that such
leaves are unifoliolate, in others the evidence is doubtful, and in yet others there is
no sign at all that the leaves are not truly simple. Even in species which sometimes
produce compound leaves the simple leaves may show no signs of being unifoliolate.
Careful anatomical investigation might provide evidence that apparently simple
leaves are really unifoliolate, but these investigations have not been carried out and
would, in any case, not be possible in practice as an aid to identification. It therefore
seems better at present to use the terms "simple" and "unifoliolate" loosely as
equivalents and not in any exact sense. The species in which simple (unifoliolate)
leaves occur are listed below; numbers in brackets refer to species in which both
simple and compound leaves occur; numbers in italics refer to species whose leaves
show clear signs of being unifoliolate. **Acanthonotus** sp. 1–3, **§Latestipulatae**
spp. *1*, 8; **§Paniculatae** spp. 1–3, (12), 16, (17); **§§Brevi-erectae** spp. 1, 2, 6, 7, 8;
§§Dissitiflorae sp. (7); **§§Brevipatentes** spp. (*1*), (5), 6, 7; **§§Pilosae** spp. *1*, 2–4;
§§Atratae spp. 1–3; **§§Geanthae** sp. (*2*); **§§Alternifoliolae** sp. (*27*); **§§Simplices-
reflexae** spp. *1–4*; **Indigastrum** sp. *6*; **Microcharis** spp. 1–8, *9–11*, (12).

(E) 1–2-seeded pods

The 1-seeded pods of *I. linifolia* (L.f.) Retz led Desvaux in 1813 to separate it
from *Indigofera* as the type of a new genus *Sphaeridiophorum* which Bentham in the
Genera Plantarum treated as a subgenus. Subsequently E. G. Baker included
15 further species having 1–2-seeded pods in his section *Sphaeridiophora*. These
species are very diverse and there seems to be no clear line between consistently
1-seeded species, such as *I. linifolia*, 2-seeded species often 1-seeded by abortion,
and 3–4-seeded species often 1- or 2-seeded by abortion. *Sphaeridiophora* has here
been dropped completely as an infra-generic taxon, as it is considered that 1–2-seeded
species have arisen independently in quite different sections of the genus, as listed
below. **Acanthonotus** all spp.; **Amecarpus** sp. 22; **§Paniculatae** spp. 4–10, 16,
17; **§§Brevi-erectae** spp. 1, 2, 12, 13, 14; **§§Anomalae**; **§§Tinctoriae** spp. 21, 22;
§§Microcarpae.

SYNOPSIS OF CLASSIFICATION

Genus	Subgenus	Section	Notes	Subsection	No. of species
CYAMOPSIS*					3*
INDIGOFERA					
	A. **Acanthonotus***				3*
	B. **Amecarpus***	1. *Amecarpus*			21*
		2. *Demissae*			3*
	C. **Indigofera**				
		1. *Latesti-pulatae**			11*
		2. *Paniculatae*		a. Paniculatae	18
				b. Trichopodae	3
		3. *Indigofera*		a. Juncifoliae	1
			Fruiting pedicel mainly erect	b. Brevi-erectae	15
				c. Anomalae*	1*
				d. Dissitiflorae	26
				e. Spinosae*	4*
				f. Brevi-patentes	7
				g. Pilosae	8
			Fruiting pedicel mainly spreading	h. Viscosae	18
				i. Centrae	12
				j. Atratae	15
				k. Psiloceratiae	11
				l. Geanthae	2
			Fruiting pedicel mainly reflexed	m. Tinctoriae	32
				n. Hirsutae*	4*
				o. Microcarpae	1
				p. Alterni-foliolae	28
				q. Simplices-reflexae*	4*
	D. **Indigastrum***				7*
	E. **Microcharis***				23*
RHYNCHOTROPIS*					2*
					283

Taxa marked * are believed to have been treated in their entirety in this paper, and the corresponding number of species is thus complete. Other taxa are thought to include further species from other geographical areas besides those treated here.

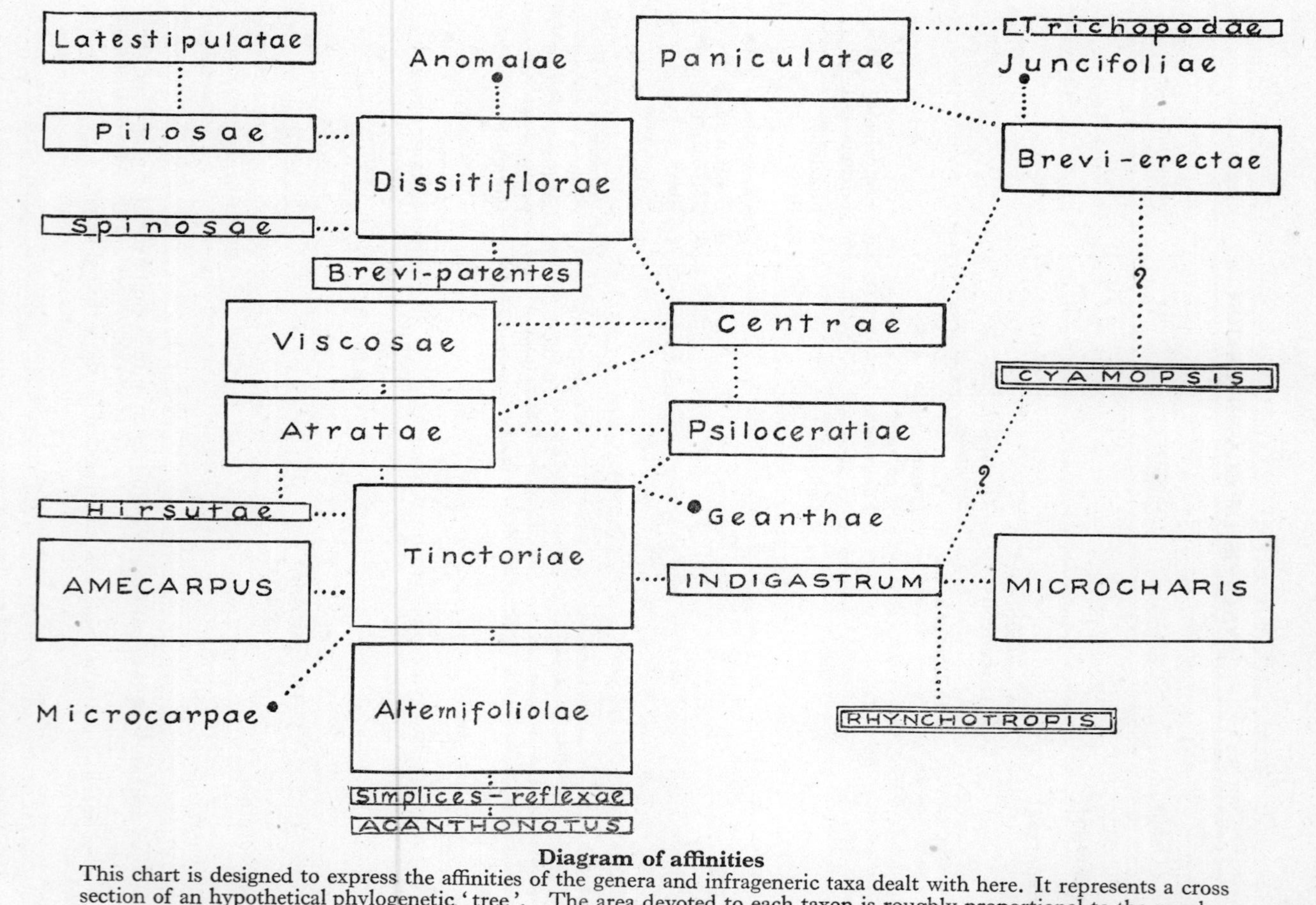

Diagram of affinities

This chart is designed to express the affinities of the genera and infrageneric taxa dealt with here. It represents a cross section of an hypothetical phylogenetic 'tree'. The area devoted to each taxon is roughly proportional to the number of species involved.

(F) Chromosome numbers

Darlington and Ammal in their "Chromosome Atlas of cultivated plants" (1945), refer to chromosome counts in 12 species of *Indigofera*, 7 of which are considered in the present paper, the remainder being asiatic species listed below in brackets.

Of these 12 species 10 belong to a polyploid series with $n = 8$.

I. aspera, I. diphylla, (*I. kirilowii*), (*I. pseudotinctoria*), *I. colutea* (*viscosa*).

diploid ($2n = 16$).

tetraploid ($2 \times 2n = 32$) *I. sessiliflora, I. suffruticosa.*

hexaploid ($3 \times 2n = 48$) (*I. decora*), (*I. divaricata?*), (*I. gerrardiana*).

Except for *I. aspera* in subgenus **Amecarpus** all these species belong to §**Indigofera.** It is interesting that $2n = 14$ is given for *I. parviflora* belonging to subgenus **Indigastrum** which is certainly far more distinct from subgen. **Indigofera** than is **Amecarpus.** *Cyamopsis tetragonoloba* (*psoraloides*) is also said to have $2n = 14$.

On the other hand the remaining count, $2n = 12$ for "*I. anil* (*sumatrana*)" published by H. A. Senn, Bibl. Genet. **12**, 175 (1938) is surprising. In the present paper *I. sumatrana* is considered to be a form of *I. tinctoria* while *I. anil* is a synonym for *I. suffruticosa.* It has not been possible to discover the species on which this count was actually made but it must have belonged to §§*Tinctoriae* and would thus be expected to belong to the $n = 8$ polyploid series.

Key to Genera and Subgenera

Vexillar filament lightly attached to its neighbours; corolla (or at any rate the standard) glabrous, veined; fruits broad, flat, longitudinally ridged, stiffly erect; leaflets 1–7, sometimes dentate or lobed . . . . **Cyamopsis**

Vexillar filament free; veins of corolla rarely apparent; leaflets never dentate or lobed:

Fruit indehiscent, 1-seeded, depressed falcate, subtriquetrous, on a reflexed pedicel; leaves simple; standard strigulose, the keel rostrate and laterally spurred; all stamens fertile, the anthers with hairs but without scales at the base . . . . A. **Acanthonotus**

Fruit not as above:

Standard nearly always $\pm$ pubescent, strigulose, or strigose outside (sometimes at the tip only), keel rarely rostrate, nearly always spurred laterally; stamens all fertile, the anthers sometimes having hairs but no scale at the base, fruit 1-many-seeded:

Pod flat, inflorescence always a raceme, fruiting pedicels reflexed; leaves pinnate with opposite leaflets (very rarely some leaves simple); bracts sometimes persistent; brown, black or glandular hairs never present; stipules never half as large as a leaflet; endocarp never dark spotted . . . B. **Amecarpus**

Pod usually $\pm$ terete; inflorescence, pedicels and leaves various; bracts not persistent; brown black or glandular hairs sometimes present; endocarp often spotted . . . . C. **Indigofera**

Standard glabrous; keel nearly always rostrate, not laterally spurred; inflorescence a raceme; brown, black or glandular hairs never present; stipules never approaching the size of a leaflet; pod often somewhat flattened, more than 2-seeded, the endocarp never spotted:

Style not twisted; apical and basal anther appendages, where present, entire, or the basal forked at the tip, never plumose:

Bracts not persistent; fruiting pedicel rarely over 2 mm. long, usually

5

reflexed, like the fruit; all stamens fertile, the anthers of the shorter
  stamens with, or more often without, basal scales: inflorescence dense
  in bud, opening later; hairs almost always appressed . . . .
  . . . . D. **Indigastrum**

Bracts persistent; fruiting pedicel often over 2 mm. long, spreading, the
  fruit itself often at right angles to it; dorsal stamen nearly always
  without an anther; anthers, at least those of the 4 shorter stamens,
  having hyaline scales at the base; inflorescence lax, hairs often
  spreading at the tips . . . . E. **Microcharis**

Style twisted, cymbiform-dilated in its lower half and abruptly contracted
  at the base where it is eventually bent through a right angle, apical and
  basal anther appendages plumose, all stamens fertile; leaves simple,
  linear or lanceolate . . . . **Rhynchotropis**

CYAMOPSIS

CYAMOPSIS DC., Mém. Lég. 230 (1825); A. Chev., Rev. Bot. Appl. **19**, 242–249
(1929).

Type species *C. tetragonoloba* (L.) Taub.

Centre of origin? South Tropical Africa?

Annuals. Hairs biramous as in *Indigofera*. Leaflets 1–7, often dentate. Racemes
sessile or shortly pedunculate in the leaf axils, fruiting pedicels stiffly erect. Corolla
glabrous (or with a few stiff hairs on the commissure of the keel in *C. serrata*),
markedly veined. Vexillum somewhat pointed. Carina not rostrate; usually merely
somewhat gibbous at the sides but shortly spurred in *C. serrata*. Stamens all fertile,
the filaments united to form a tube, but the vexillar stamen less strongly attached
to its neighbours than the others. Anthers apiculate, without basal scales. Fruit
broad, flattened, with 3 ridges on each face, many-seeded, ending in a pronounced
beak. Seeds flattened, tuberculate.

A very natural group undoubtedly closely related to *Indigofera*. The characters
of *C. serrata*, e.g. its spurred carina, to some extent break down the distinction
between the two genera but the number of characters involved, even though each
by itself is small, the fact that there is no group of species within *Indigofera* which
is not readily separable from *Cyamopsis*, the undesirability of changing the name
of *C. tetragonoloba*, which is an important economic plant, and the general aspect
of this species, which, as may appear from its synonyms, is quite different from that
of any species of *Indigofera*, make it desirable to retain *Cyamopsis* as a distinct genus.

Leaflets, of later leaves, 3, always toothed, the length of the teeth less than $\frac{1}{10}$ of the
  breadth of the leaflet, which usually exceeds 1 cm.; carina somewhat gibbous
  at the sides, not spurred; hairs on fruit appressed . . . . . 1. *C. tetragonoloba*.

Leaflets, of later leaves either entire or with large teeth whose length is usually more
  than $\frac{1}{10}$ of the breadth of the leaflet, which rarely exceeds 1 cm.:

Later leaves 5–7-foliolate; carina somewhat gibbous at the sides, not spurred:

Leaflets entire; axis of raceme usually over 2 cm. long; hairs on fruit sparse
  and appressed . . . . 2. *C. senegalensis*.

Leaflets sometimes toothed or lobed; raceme short or long; hairs on fruit
  (where seen) numerous and spreading or appressed . . . .
  . . . . Between 2 and 3. *intermediate forms*.

Later leaves 3-foliolate, leaflets entire or, deeply lobed; carina shortly spurred;
  axis of raceme under 1 cm. long; hairs on fruit dense and spreading . . . . .
  . . . . 3. *C. serrata*.

1. **C. tetragonoloba** (*L*) *Taub*. in Engl. & Prantl. Pflanzenfam. **3**: **3**, 259 ; Chev. 1.c. 244, t. **2** A.

Psoralea tetragonoloba L., Mantissa 104 (1767).
Dolichos fabaeformis L'Hér., Stirp. Nov., t. 78 (1784).
D. psoraloides Lam., Encyc. **2**, 300 (1786).
Lupinus trifoliolatus Cav., Ic. Pl. Rar. **1**, t. 59 (1791).
? Galega esculenta Rottboll ex Spreng., Nachr.: Bott. Gart. Halle 23?
Cyamopsis psoraloides (Lam.) DC., Mém. Lég. 231 (1825).

Apparently unknown in the wild state: its cultivation is ancient in India where several varieties are known. During the last three centuries it has been introduced into many other tropical countries.

2. **C. senegalensis** *Guill. & Perr.*, Fl. Seneg. 171, t. 45 (1832); L.T.A. p. 94; Chev. 1.c. 246, t. 2. B. excl. syn.

C. senegalensis var. *stenophylla* Bonnet, Bull. Soc. Bot. Fr. **58**, mem. **20, 6** (1911).
C. stenophylla (Bonnet) Chev., 1.c. 246, t. **2**. C. (1939).

ARABIA, nr. Jeddah *Schimper* (1857) 840, *Zohrab* 118, 184, *Fischer* 67, *Trott* 50, 1296, 1371, 1542, *Bates* 2140 BM; SENEGAL (*Perrottet* s.n. P. holotype); FR. SUDAN Timbuktu distr. *Hagerup* 280 BM (*Chudeau* 20 P type of *C. stenophylla*); A.-E. SUDAN, Suakin-Berber *Schweinfurth* 677, K. BM; Gash delta Kassala Prov. *Andrews* 242; Mt. Arasch Kool *Kotschy* (1839) 141, K, BM. Gedary distr. *Baliker Beshir* 48; ERITREA Beni Amer Carajai *Pappi* 6256 K, BM.

Chevalier's *C. stenophylla* appears to be nothing but a narrow-leaved form, growing in more arid conditions, and passing gradually into the more typical form of the species.

3. **C. serrata** *Schinz*, Abh. Bot. Ver. Brand. **30**, 161 (1887); L.T.A. p. 95.
Indigofera psammotropha Bolus, J. Bot. **34**, 22 (1896).

S.W. AFRICA, Amboland, Oshiheke, *c.* 18°S, *Schinz* s.n. Z holotype; Great Namaqualand *Pearson* 4346, 4481, 4679, 4875, 4950, 4879. CAPE PROV., Kimberley distr. *Steyn* 2727; Hay distr. *Steyn* s.n. *Wilman* 2267, 2275; Prieska distr. *Wilman* s.n., *Bryant* 1137; Hopetown distr. (*Muskett* in BOL 2039 type of *I. psammotropha*).

On a single plant some leaflets may be quite entire while others are deeply toothed or pinnatifid.

3a. forms intermediate between *C. serrata* and *C. senegalensis*.

ANGOLA, river Caruryamba *J. B. Texeira* 529-A BM; TROPICAL BECHUANALAND, Kwebe Hills *Lugard* 128, 135 syntypes of *Indigofera dentata* N. E. Brown, Kew Bull. **1909**, 102; L.T.A. p. 144. S.W. Africa Ameib *Dinter* 6864, Okahandja, 1300 m., *Dinter* 469 K, BM, Kuiseb River, 600 m., *Mrs. Macdonald* 491 BM.

These plants vary among themselves; they have 3–7-foliolate leaves as in *C. senegalensis* but with rather wider leaflets. A flower of *Lugard* 135 had a carina merely gibbous at the sides as in *C. senegalensis* not spurred as in *C. serrata*. On the other hand all 6 specimens have leaflets sometimes toothed as in *C. serrata*. *Lugard* 128, 135 and *Dinter* 6864 have long inflorescences as in *C. senegalensis*. They show no fruits. *Dinter* 469 and *Mrs. Macdonald* 491 have short inflorescences and fruits with copious spreading hairs as in *C. serrata*. In *Texeira* 529-A the hairs on the fruit are short and appressed as in *C. senegalensis*.

Were these 6 sheets typical of *C. serrata* there would be much to be said for Chevalier's treatment of that species as synonymous with *C. senegalensis*. It seems likely in fact that Chevalier only saw one such sheet and did not see typical *C. serrata* at all. After taking into account the fact that the great majority of specimens of *C. serrata* differ consistently from *C. senegalensis* in the 5 characters listed in the key, while only in the north of its range do the present intermediate forms appear, it seems more satisfactory to maintain *C. serrata* as a distinct species and to leave unsettled for the moment the status of these intermediate forms, which may represent one or two definite varieties or species, but may on the other hand be the result of hybridisation between *C. serrata* and *C. senegalensis*.

INDIGOFERA L., Hort. Cliff. 487 (1737) p.p., Fl. Zeyl. 124 (1748), Gen. Pl. ed. **4**, 431 (1752), ed. **5**, 333 (1754), sp. pl. 751 (1753).

Type species *I. tinctoria* L.

Subgenus A. **Acanthonotus** (Benth.) Benth & Hook.f., Gen. Pl. **1**, 494 (1865).

Acanthonotus Benth., Fl. Nigrit. 293 (1849) (as genus).

Section *Echinatae* Wight et Arn., Prod. Fl. Pen. Ind. Or. 198 (1834); L.T.A. 96.

Type species *I. nummulariifolia* (L.) Livera ex Alston.

Leaves simple, inflorescence an axillary raceme, corolla strigulose outside, pointed in the bud. Pods, short, falcate, subtriquetrous, 1-seeded, indehiscent, the endocarp not spotted.

The closest affinity seems to be with §§*Simplices—reflexae*.

Pod echinate, leaves (in African plants) usually appressed-strigulose above; an erect or prostrate annual, not (or very rarely) rooting at the nodes . . . .
1. *nummulariifolia*

Pod, not echinate:

Leaves subcordate, glabrous above, except at the margins; prostrate, apparently often perennial and rooting at the nodes . . . . 2. *drepanocarpa*

Leaves $\pm$ cuneate at the base . . . . 3. *cuitoensis*

A.1. **I. nummulariifolia** (*L.*) *Livera ex Alston* in Trim., Handb. Fl. Ceylon **6**, suppl. 72 (1931).

Hedysarum nummularifolium L., Sp. pl. 746 (1753).

I. echinata Willd., Sp. Pl. **3**, 1222 (1803); L.T.A. No. 1; F.C.B. **5**, 125.

Acanthonotus echinatus (Willd.) Benth., Fl. Nigrit. 293 (1849).

India: Tranquebar Madras prov. (*unknown collector* s.n. B type of *I. echinata*). Ceylon: *Hermann* s.n. BM. holotype; Indo China; Senegal; Port Guinea; French Guinea: *Chevalier* 18874 BM; Fr. Sudan; Gold Coast; Nigeria; Cameroons; Fr. Eq. Africa; B. Congo; A.-E. Sudan; Tanganyika, T1, T4–6, T8; Port. E. Afr.; S. Rhodesia; N. Rhodesia; Madagascar.

Botanists have disagreed on the interpretation of *Hedysarum nummulariifolium*. Linnaeus' description in the Species Plantarum is taken verbatim from his "Flora Zeylanica" No. 288, p. 133. The descriptions in this latter work are known to be based on Hermann's plants, now preserved in the British Museum. The description of *H. nummulariifolium* agrees well with Hermann's specimen but is quite incompatible, for instance in its "Legumina declinata, lunata, dorso hispida, monosperma" with Petiver's plate Gaz. t. 26 f. 4 which, although cited by Linnaeus, following Burman, as a synonym, is a good illustration of an *Alysicarpus*. There can be no doubt that the type of the species is the plant which Linnaeus actually described, and which fits his description, and not a plate, which he may not even have seen, and cannot have taken into account in drawing up his description. In Africa the leaves are usually appressed-strigose above; in India they are usually glabrous.

A.2. **I. drepanocarpa** *Taub.* in Engl. Pflwelt. O. Afr. C, 209 (1895); L.T.A. No. 3; F.C.B. **5**, 125.

I. erythrogrammoides De Wild., Ann. Mus. Congo **5**: **1**, 133 (1904).

N. Nigeria, *Lely* 481. Oubangui-Chari, *Tisserant* 240 bis BM; B. Congo, *Hendrickx in Gillet* 3063 BR. holotype of *I. erythrogrammoides*; Uganda, U1–4; K7; Tanganyika, T1 (*Boehm* 109 B†, *Stuhlmann* 3503 B†, 4653 B† syntypes) Bukoba *Stuhlmann* 1039 neotype; T3–6; Pemba; N. Rhodesia, Abercorn distr.

Stuhlmann 1039 was determined by Taubert as "*I. drepanocarpa* Taub. sp. n." and is probably a duplicate of a plant mentioned in Taubert's description as "Bu. hfg. Stuhlm." It would therefore seem suitable as a neotype. This species varies considerably in leaf size, indumentum, etc. The type of *I. erythrogrammoides* approaches *I. cuitoensis*.

A.3. **I. cuitoensis** *Bak.f.*, L.T.A. 96 (1926) No. 2.

Angola, *Gossweiler* 2578 BM. holotype K. isotype. Doubtfully distinct from *I. drepanocarpa*.

Subgenus B. **Amecarpus** Benth. ex Harvey, Fl. Cap. **2**, 201 (1862); L.T.A. 163 (as section) lectotype species *I. sessilifolia* DC.

Centre of distribution Southern Africa.

Leaves pinnate, the rhachis prolonged beyond the lateral leaflets, which are always opposite; rarely some leaves trifoliolate, or simple on the flowering shoots only, Inflorescence an axillary raceme, fruiting pedical usually ± reflexed. Calyx teeth usually much longer than tube. Standard strigulose outside. Pod flattened, strigulose or pubescent, the septa between the seeds thin, endocarp without dark spots.

Flattened pods are also found in Subgen. *Microcharis* and in Subgen. *Indigofera* sect. *paniculatae*.

Shoots of 2 different kinds from a woody rootstock, the flowering shoots having 1–3-foliolate leaves, the vegetative shoots having 5–19-foliolate leaves . . . . .
1. *letestui*.

Shoots of one kind:

Pod 2·4 mm. wide or more; bracts rarely persistent; stamens 3–5·5 mm. long:

Small rigid shrubs:

Petiole not above 2 mm. long; axis of inflorescence ± spinescent:

Axis of inflorescence a brown spine, less than 2 cm. long, even in fruit, pod curved in a ± complete circle, the tip almost touching the base; dorsal suture of pod very prominent, with a thin vertical margin, about 0·5 mm. broad rising above it . . . . 2. *circinnata*.

Axis of inflorescence ± spinescent, usually over 2 cm. long in fruit; pod falcate, not curved into a circle no thin margin above the dorsal suture of the pod . . . . 3. *sessilifolia*.

Petiole up to 4 or 5 mm. long; axis of inflorescence not spinescent; fruits curved into a circle; dorsal stuture of pod prominent, having a very narrow margin, *c*.0·2 mm. wide, above it . . . . 4. *rautanenii*.

Herbaceous, or somewhat woody at base only:

Calyx more than ⅓ of length of standard:

Leaflets 3–7, if 7 then fruiting inflorescence as long as, or longer than, subtending leaf; fruit usually having a lateral ridge, or bosses over the seeds:

Pod usually ± curved, with conspicuous prominent veins; surface of seeds ridged; stamens 3–4 mm. long; terminal leaflet up to 40 mm. long. . . . . .5. *rhytidocarpa*.

Leaflets 1–3, petiole usually under 8 mm. . . . . . . α ssp. *rhytidocarpa*.

Leaflets 3–7, petiole 8–13 mm. . . . . β ssp. *angolensis*.

Pod without conspicuous veins, or, if a few in No. 7, then stamens *c*.5 mm. long:

Pod dark red-brown, 1–2-seeded, with sparse short, appressed hairs *c*.0·2 mm. long, stamens *c*.3 mm. long . . . . . . .6. *leendertziae.*

Pod brownish, straw coloured or silvery with longer hairs and usually more seeds:

Filaments *c*.5 mm. long:

Calyx definitely shorter than filaments; leaflets up to 3 times as long as wide; pod with separate rounded eminences over each seed, not united into a continuous lateral ridge; style base in fruit upturned; seeds 5–8 . . . . .7. *holubii.*

Calyx about as long as filaments; leaflets up to 8 times as long as wide; pod with a continuous lateral ridge which is weakened but not interrupted between the seeds; style base pointing straight forward or slightly downward; seeds 3–5
18. α. *charlieriana* var. *lata.*

Filaments 4 mm. long, or less:

Perennial, hairs on pod closely appressed; indumentum on stem dense:

Hairs on pod up to 0·5 mm. long . . . . .8. *torulosa.*

Leaflets ± elliptic oblong . . . . α. var. *torulosa.* .

Leaflets ± lanceolate, acute . . . . β. var. *angustiloba.*

Some of the hairs on pod up to 1·0 mm. long . . . 10. *arabica.*

Annual:

Fruit 1–2-seeded with long spreading hairs, usually 3 mm. or more broad . . . . .11. *chirensis.*

Fruit more than 2-seeded, often less than 3 mm. broad:

Fruit ± curved, 5–9-seeded, usually *c*.2·5 mm. broad; fruiting peduncle, not much more than half as long as a fruit . . . . .12. *hochstetteri.*

Fruit ± straight, 3–6-seeded; fruiting peduncle 10 mm. or more long, as long as, or longer than a fruit:

Calyx more than half as long as stamens; leaves sparsely strigose above; terminal leaflet 7–9 times as long as wide . . . . 13. *senegalensis.*

Calyx about half as long as stamens; leaves glabrous above; terminal leaflet 2–4 times as long as wide . . . . . .
14. *praticola.*

Leaflets 7–11; fruiting inflorescence shorter than leaf; pods slightly curved, without a lateral ridge, or eminences above the 5-8 seeds . . . . .16. *wilmaniae.*

Calyx less than a third of length of standard and stamens . . . 17. *dauensis.*

Pod 2·2 mm. wide, or less:

Stamens 6·5–9 mm. long; bracts caducous; seeds not fewer than 8:

Stems angled, sometimes flattened, 2–3 mm. thick at the bases of the inflorescences; stipules up to 8 mm. long; stipellae well developed; leaflets up to 4 cm. long and 12 mm. wide; hairs at margin of leaflet resembling those of its under surface, larger and stouter than those of the upper surface; inflorescences up to 40 cm. long; stamens *c*.9 mm. long . . . . 18. *lasiantha.*

Stems cylindrical, *c.*1 mm. thick at the bases of the inflorescences; stipules up to 2 mm. long; stipellae hardly visible; leaflets up to 2·5 cm. long and 6 mm. wide; hairs at leaflet margin resembling those of the upper surface; inflorescence up to 28 cm. long; stamens *c.*7 mm. long . . . . .
19. *fanshawei.*

Stamens 2–5·5 mm. long:
Bracts caducous; seeds rarely fewer than 8, or, if 4·6, then pod 2 mm. wide:
Pod 2 mm. wide, obtuse-truncate, the end (apart from the style base) being ± at right angles to the upper edge; lateral ridge ± interrupted:
Annual; hairs on fruit and stems sparse; leaflets glabrous above; pods ± straight *c.* 8-seeded . . . . 15. *sisalis.*
Perennial; stems fruits and leaves closely covered with appressed white hairs; pods ± curved, *c.* 4–6-seeded . . . . 9. *oligophylla.*
Pod acute, the end forming an angle of *c.*45° with the upper edge; annuals;
Pod 1·6–2 mm. wide, the lateral ridge ± continuous . . . 20. *charlieriana.*
Inflorescence as long as or longer than the leaves; leaves 1–5 jugate
α. var. *charlieriana.*

Inflorescence shorter than the leaves; leaves 1–2 rarely 3 jugate; leaflets rarely over 3 mm. wide:
Pod densely covered with hairs *c.*1·0 mm. long . . β. var. *scaberrima.*
Pod sparsely covered with hairs *c.*0·5 mm. long . . . γ. var. *sessilis.*
Pod 1–1·5 mm. wide, the lateral ridge broken into swellings above the seeds, fruiting inflorescence usually more than twice as long as subtending leaf, wider leaflets 4–8 mm. wide . . . . . . 21. *aspera.*
Bracts persistent; seeds usually fewer than 8, if more than 4, then pods less than 1·5 mm. wide
§**Demissae.**
Fruit 1–2-seeded, up to twice as long as wide; leaflets usually 5–7, up to 7 mm.; calyx and stamens *c.*2 mm. long . . . . 22. *demissa.*
Fruit *c.* 5-seeded, 3 or more times as long as wide; leaflets usually 9:
Leaflets up to 7 mm.; calyx and stamens *c.*2 mm. long . . 23. *concinna.*
Leaflets up to 13 mm.; calyx *c.*3·5 mm. stamens *c.*4 mm. long . . . .
24. *bussei.*

B.1. **I. letestui** *Tiss.*, Bull. Mus. Hist. Nat. Par. 2nd ser. **3**, 171 (1931).

I. mittuensis Bak.f., J. Bot. **70**, 252 (1932).

OUBANGUI-CHARI (*Chevalier* 7308) (*Le Testu* 3650 P), *Tisserant* 950 P (950 bis P) (950 ter P) syntypes; A.E. SUDAN, Mittu land *Schweinfurth* 2777 holotype of *I. mittuensis.*

B.2. **I. circinnata** *Benth. ex Harv.*, Fl. Cap. **2**, 202 (1862).

S. RHODESIA. Gwanda distr., *R. M. Davies* 892; EXTRA TROP. BECHUANALAND, *Harbor in Rogers* 4377; TRANSVAAL, *Burke* 355 holotype, etc.

B.3. **I. sessilifolia** *DC.*, Prod. **2**, 231 (1825); Fl. Cap. **2**, No. 111.

I. patens Eckl. & Zeyh., Enum. Pl. Afr. Austr. 236 (1836); Fl. Cap. **2**, No. 108.
I. falcata E. Mey., Comm. Pl. Afr. Austr. 93 (1836); Fl. Cap. **2**, No. 109, probably *I. hedranophylla* Eckl. & Zeyh., Enum. 232 (1836), Fl. Cap. **2**, No. 110 (not seen).

I. melolobioides Benth. nomen.

S.W. Afr., *Pearson* 8209, etc.; Orange Free State, *Verdoorn* 1635, etc.; Central, N. Central and East Cape, *Burchell* 1746 isotype, *Drége* s.n. near Plat drift isosyntype of *I. falcata* (*Ecklon & Zeyher* 1580 type of *I. patens*, 1560 type of *I. hedranophylla*) *Burke & Zeyher* 499, etc.

There is some variation in petiole length, indumentum, size of leaflets and length of inflorescence, but there seem to be no discontinuities sufficient to warrant the continued subdivision of the species.

B.4. **I. rautanenii** *Bak.f.*, Vierteljahrschr. Nat. Ges. Zur. **49**, 189 (1904); L.T.A. No. 248.

S.W. Afr., *Rautanen* 464 isotype, *Bradfield* 61, *Dinter* 7785.

B.5. **I. rhytidocarpa** *Benth. ex Harv.*, Fl. Cap. **2**, 202 (1862); L.T.A. No. 251. ssp. **rhytidocarpa.**

S. Rhodesia, Plumtree-Embakwe, Bulalime-Mangwe distr. *Feiertag* in SRGH 45369 SRGH; Transvaal, *Schlechter* 11784, etc.; Natal, *Wood* 8255, *Gerrard* 1713; Orange Free State, *Burke & Zeyher* s.n. holotype, etc.

β ssp. **angolensis** *Gillett* ssp. nov.

"*I. anabaptista* Steud." sensu Hiern, Cat. Welw. Afr. Pl. **1**, 217; Bak.f., J. Bot. **41**, 332 (1903) p.p.; L.T.A. 164 p.p. quoad pl. angolenses, non. Steud.

Herba annua quasimodo inter *I. hochstetteri* et *I. rhytidocarpam* intermedia, sed fructu *I. rhytidocarpae*. Ab ssp. *rhytidocarpa* foliolis 3–7 (non semper 3) brevioribus, petiolis longioribus (8–13 mm. nec 4–8 mm.), fructibusque minus rufescentibus differt. Semina 4–6.

Angola, Loanda distr. dry and moist sandy places near Morro das Lagostas, Boa Vista and Praia de Zamba grande *Welwitsch* 4137 K, holotype, BM, isotype, 4138; K, BM, 4139; K, BM, 4140; K, BM, 4141, K, BM; on hill slopes facing the sea S. of Loanda, after the rains *Gossweiler* 130, 182 BM, 250, 258 BM, 265, 328 BM, 781 BM, 5878 BM; Mossamedes distr. *Welwitsch* 4142 K, BM; Huilla, Chianje *N.M. Pritchard* 382 BM; S.W. Afr. Dammaraland *Een.* s.n.BM.

B.6. **I. leendertziae** *N.E. Br.* in Burtt Davy, Man. Flor. Transvaal **2**, XXIV & 367 (1932).

Transvaal, Potgietersrust, *Leendertz* 1942 holotype.

This may prove but a form of *I. torulosa*.

B.7. **I. holubii** *N.E. Br.*, Kew Bull. **1925**, 148.

S. Rhodesia, Ndanga 600 m. *Bates* in SRGH 16901; S.W. Afr., Waterberg *Bradfield* 340; Transvaal, *Holub* s.n.type, *Codd* 6625, *Pott* 4223; Extra trop. Bechuanaland, *Holub* s.n. syntype; N. Central Cape, *Bryant* 1067, *M. Wilman* 3106.

B.8. **I. torulosa** *E. Mey*, Com. Pl. Afr. Austr. 105 (1836); Fl. Cap. **2**, No. 113. α. var. **torulosa.**

Natal, *Gerrard* 1094; E. Cape, *Drége* s.n. isotype, *Galpin* 5862, *Pegler* 1805 K, BM, etc.

β. var. **angustiloba** (*Bak.f.*) *Gillett* stat. nov.

I. angustiloba Bak.f., Vierteljahrschr, Nat, Ges, Zur. **49**, 183 (1904).

"*I. torulosa E. Mey*" sensu N.E. Br. in Burtt Davy. Man. Fl. Transvaal **2**, 367 non E. Mey.

Transvaal, *Wilms* 320 syntype, *Galpin* 12189, *Rogers* 14559, *Schlechter* 3948, 6387, *Louw* 1664, *Pott* 3834.

B.9. **I. oligophylla** *Klotzsch*, Pet. Moss. Bot. 49 (1862); L.T.A. No. 250.

PORT. E. AFR., Island of Goa (*Peters* s.n. B † holotype); same locality 15°3′8″S × 40°47′8″E, in the bush on coral and sand, flower violet, common, 5.5.1947, *Gomes e Souza* 3512 K, neotype, PRE, SRGH isoneotypes. Island of Moçambique, junto to Fortaleza, prostrate perennial, 31.10.1942, *Mendonça* 1160 K, LISC.

Most closely related to *I. torulosa* var. *angustiloba* from which it differs, apart from the slightly narrower fruits, in the longer petioles and leaflets, a more silky indumentum, shorter stipules and longer stamens (*c.*4·5 instead of 3–3·5 mm.).

B.10. **I. arabica** *Jaub. & Spach*, I11. Pl. Or. t. 479 (1856); non sensu Bak., F.T.A. **2**, 103.

YEMEN (*Botta* s.n. P, holotype); ADEN *Dr Lunz* 21, *Hunter* 230, *Beevor* 62, *Hooker* 66; ERITREA, *Bally* 6800; ETHIOPIA, Aubarre, 43°13′E × 9°48′N, *Hummel* 16 EA; BRIT. SOMALILAND, Rahale, 42°52′ × 10°59′, *Godding* 210, Buramo, 43°10′ × 10°2′, *Gillett* 4915, Debrawen, 42°49′ × 10°26′, *Gillett* 4653.

This material differs from Jaubert & Spach's plate in having leaves 3–5- instead of 5–7-foliolate, and the style base frequently asymmetric and pointed upwards instead of central and pointing forward in the fruit. Schwarz in Mitt. Inst. Bot. Hamb. **10**, 105 (1939) gives a much wider distribution of this species in Arabia from Sinai to Dhofar, but specimens have not been seen.

B.11. **I. chirensis** *Cufod.*, B. J. Bot. Ét. Brux. **25**, 263 (1955).

I. ervoides A. Rich., Tent. Fl. Ab. **1**, 179 (1847); L.T.A. No. 252; non *Meisn.* (1844).

ERITREA, *Pappi* 286 (352) K, BM; N. ETHIOPIA, *Quartin Dillon* s.n. (P, holo.) K, isotype, *Schimper* (1863–8) 1545. K, BM.

B.12. **I. hochstetteri** *Bak.*, F.T.A. **2**, 101 (1871); F.C.B. **5**, 150.

I. anabaptista Steud, ex Bak. in Hook.f., Fl. Brit. Ind. **2**, 102 (1876); L.T.A. No. 249.

I. ornithopodioides Hochst. & Steud. ex Jaub. & Spach, Ill. Pl. Or. t. 480 (1856), non Schum. et Thonn. (1829) nec Cham. et Schlecht.

I. jaubertiana Schwnf., Bull. Herb. Boiss. **4** app. **2**, 245 (1876).

I. arenaria A. Rich., Tent FL. Ab. **1**, 183 (1847); non E. Mey. (1836).

"*I. arenaria* A. Rich." sensu Bak., F.T.A. **2**, 79; Bak.f., J. Bot. **41**, 192; L.T.A. No. 44; Andrews, Fl. Sudan **2**, 208 et *I. semhaensis* Vierhap, Denk. Ak. Wien. Math. Nat. **71**, 362 (1907) pro minore parte, quoad typum solum.

HEJAZ; YEMEN, *Schimper* (1837) 769 isotype of *I. hochstetteri*, *I. anabaptista*, *I. ornithopodioides* and *I. jaubertiana*; ADEN; PUNJAB; SIND; CENTRAL INDIA; MADRAS; FR. SUDAN, Timbuktu *Hagerup* s.n. BM; FR. NIGER COL., Aïr, 17°45′N., *Chopard et Villiers* s.n. IFAN; NIGERIA, *Meikle* 1097, *Lely* 621; B. CONGO (fide Cronquist); A.E. SUDAN; ERITREA, *Dillon et Petit* s.n. P, holotype of *I. arenaria*, *A. Rich* and *I. semhaensis*; ETHIOPIA; BRIT. SOMALILAND, *White* 165; UGANDA, U2; KENYA, K2–4; TANGANYIKA, T1, T2, T4, T7 *Maclunes* 284 BM.

The type of *I. arenaria* A. Rich is a very young plant without fruits. Until now it does not seem to have been studied by authors subsequent to Richard and his name has been universally misapplied to a totally different species (*I. disjuncta*).

B.13. **I. senegalensis** *Lam.*, Encyc. **3**, 248 (1789); L.T.A. No. 253.

Brissonia trapezicarpa Desv., J. Bot. **1814, 1**, 78 (1814).

SENEGAL, *Perrottet* s.n., *Heudelot* s.n. (*Geoffroi* s.n. P, holotype). *Unknown collector* s.n. P, holotype of *Brissonia trapezicarpa*; FR. SUDAN, *Chevalier* 24906; N. NIGERIA, *Moiser* 252, Bornu *Oudney* 6 BM, FR. CHAD. TERR., *Chevalier* 9743; *Talbot* 523 BM.

In his original description of *B. trapezicarpa* in 1814 Desvaux writes "habitat in India", a country where *I. senegalensis* is not known, although the related *I. hochstetteri*, which was at first confused with *senegalensis* is found. In 1826 in Ann. Sci. Nat. 1st ser. **9**, 411 Desvaux himself says that *B. trapezicarpa* is identical with *I. senegalensis*. Thanks to the kindness of the Director of the Herbarium of the Muséum d'Histoire Naturelle in Paris it has been possible to examine Desvaux's type, which is certainly *I. senegalensis* and not *I. hochstetteri*. Someone, probably Desvaux himself, annotating the sheet, apparently in 1826, has written "habitat in Senegalia" at the bottom, so that it would seem that Desvaux altered his first opinion as to the plant's origin. A confusion between the two countries could in those days readily arise, through a returning traveller adding a few plants in Senegal to a collection made in India.

B.14. **I. praticola** *Bak.f.*, Vierteljahrschr. Nat. Ges. Zur. **70**, 218 (1925) fig. 2; L.T.A. No. 255.

TANGANYIKA, T8 *Busse* 2769 EA, *Evans* 19 EA; NYASALAND, *Jackson* 1682; S. RHODESIA, *Holub* s.n. Z holotype, BM fragment, *Whittall* in SRGH 9075, *Williams* in SRGH 11827, *Robinson* 330; N. RHODESIA, *Rogers* 8731, 26000, *Robinson* 858; EXTRA TROP. BECHUANALAND, *Rogers* 6592.

This species is very close to *I. senegalensis* and some of the material listed above has previously been identified as that species.

B.15. **I. sisalis** *Gillett* sp. nov.

I. praticolae Bak.f. et *I. hochstetteri* Bak. affinis. Ab prima legumine longiori, angustiori; ab secunda legumine subrecto, angustiori, corolla staminibusque in ratione calycis multo longioribus, foliolis superne semper glabratis differt.

Herba, ut videtur, annua, ramis patentibus vel prostratis, sicut foliis, inflorescentiis, calycibusque sparse appresse strigosis, pilis 0·6–0·8 mm. longis. *Folia* 2–3-jugata in toto ad 5 cm. longa, rhachide ultra foliola lateralia opposita prolongata, stipulis subulatis ad 5 mm. longis, petiolo 10–15 mm. longo, stipellis filiformibus ad 1 mm. longis, petiolulis *c*.1 mm. longis, foliolis anguste ellipticis,· basin attenuatis acutis, apice rotundatis, superne glabris, ad 16 mm. longis et 6 mm. latis, terminale aliis subsimile. *Racemi* post anthesin ad 9 cm. attingentes, pedunculo ad 12 mm. longo incluso, bracteis anguste lanceolatis, strigosis, mox deciduis, *c*.1·5 mm. longis, pedicellis post anthesin valde reflexis. *Calyx*, tubo 0·5 mm. longo, stramineo, perlucido, nervis 5 atro-brunneis in dentes angustissimos, subaequales, 1·5–2·5 mm. longos, euntibus, in toto quam stamina duplo vel triplo brevior. *Corolla* mox decidua. *Vexillum* extus strigosum, oblongo-obovatum, apice rotundatum, 4 mm. longum, 2·5 mm. latum. *Alae.* glabrae, ± oblongae, brevissime unguiculatae, 3·5 mm. longae. *Carina* ad margines et ad commissuram strigosa, alius glabra, calcarata, haud rostrata, *c*.4 mm. longa. *Stamina* omnia fertilia, *c*.4 mm. longa, antheris basin muticis, 0·5 mm. longis, mucrone apicale brunnea 0·50 mm. longa inclusa. *Ovarium* strigosum, stylo cylindrico, base except a, glabro, medio rectangule curvato, stigmate capitato stylo duplo latiore. *Legumen* stramineum, subrectum, valde complanatum, margine incrassatum, ad semina umbonatum (vel interrupte longitudinale costatum), inter semina paulo transverse costatum, apice rotundato truncato (margine inferiore valde curvato marginem superiorem paulo curvatum ad angulum *c*.90° attingente) 12–18 mm. longum, 2 mm. latum, 0·5 mm. crassum. *Semina* 7–10, brunnea, quadrata, 1·2 mm. longa, 1 mm. lata, 0·4 mm. crassa.

KENYA, K7, Taveta estates, 750 m., annual on cultivated Sisal land 17.5.1934, *H. C. Sampson* 97 K, EA. TANGANYIKA, T2, Mwika near Kilimanjaro 900 m., creeping in sandy open spots in savannah, 20.6.1932, *C. Harvey* 9 K holo., EA isotype. T3, Lushoto distr. Mazinde, 435 m., roadside and cultivated ground, spreading prostrate herb, corolla pale red, 1.5.1953, *Drummond & Hemsley* 2339. T6, Morogoro, 840 m., May 1930, *A. S. Haarer* 1801 K, EA. West of Morogoro

on right bank of the Ngerengere river, 560 m., annual, spreading, fls. red, 29.3.1926,
A. Peter 39278; North west side Uluguru mts., 800 m., herb in clusters among
grass and on roadsides, 2.5.1933, *Schlieben* 3870 BM, 40 Km. from Morogoro on
Dar-es-Salaam road, 400 m., pioneer on bare ground beside road in *Pseudoberlinia-*
Combretum woodland, prostrate. Flowers salmon pink, 8.2.1955, *J. R. Welch* 272.
A. C. Brooks 46 from the central Serengeti plains 1350 m. may also be this species.
It is a seedling, with no fruits and is atypical in having 3-foliolate leaves which are
not glabrous above. From a sketch preserved in the British Museum it seems that
this species may have been provisionally given the m.s. name *I. stuhlmannii* by
Taubert.

B.16. **I. wilmaniae** *Bak.f. ex Gillett* sp. nov.

M. Wilman, Checklist Flow. Pl. Griqualand West 64 (1946), nomen.

Herba annua, ramis patentibus ad 25 cm. longis, sicut foliis, inflorescentiis calycibusque, viridibus, pilis rigidis appressis ad 1·0 mm. longis haud dense vestitis. *Folia*
[1-]-3-4-jugata, ad 4·5 cm. longa, rhachide ultra foliola lateralia opposita prolongata,
stipulis petiolo 1 mm. connatis, basin deltoideis, sursum subulatis, in tota ad 3 mm.
longis, petiolo 6–10 mm. longo, stipellis nullis, petiolulis 1 mm. longis, foliolis
aequalibus (terminale aliis haud maiore) anguste oblongis, apice rotundatis, basin
acutis, *c.*14 mm. longis, 4 mm. latis, superne viridioribus, pilis angustioribus.
Racemi 15–20-flori, 5–25 mm. longi, pedunculo 1–5 mm. longo incluso, pedicellis
ad 1 mm. longis, ad anthesin patentibus, post anthesin reflexis. *Calyx* strigosus,
corolla staminibusque subaequilongus [vel paulo brevior], tubo *c.*1 mm. longo,
dentibus angustis, *c.*3 mm. longis. *Vexillum* carinaque extus strigosa mox caduca.
*Stamina c.*4 mm. longa. *Legumen* paulo curvato, laeve, sine nervibus, stramineum,
valde complanatum, margine incrassato, ad 25 mm. longum 3 mm. latum, propter
seminibus paulo incrassatum, sed haud umbonatum et haud costatum, apice
inaequaliter acutum, margine inferiore rotundato marginem superiorem subrectum
ad angulum 60°–80° attingente. *Semina* 3–8, brunnea, quadrata, valde complanata,
2·5 mm. longa, 2 mm. lata. margine 0·2 mm., centro 0·5 mm. crassa.

S.W. AFR., Tiraz 64 Km N. of Aus. 1410 m, bank of dry watercourse, 5.2.1950
Mrs. Macdonald 280 BM. Between Dabaigabis and Gründoorn 1260 m., sandy
places among rocks, 3.2.1909, *H. H. W. Pearson* 3153. N. CENTRAL CAPE, Prieska
div. common all over district usually on hillsides or in kloofs, corolla pink, May
1935, *E. G. Bryant* 1139.Griqualand west; near St. Clair, Douglas on Orange
river, Dec. 1896, *K. Orpen* 132. Kimberley, Jan. 1938, *J. P. H. Acocks* 5590.
Niekerks Hoop, red sand, frequent Mar. 1921, *M. Wilman* 1384 K holo BM isotype.
Mazelsfontein, in a gully in dolomite below Kaap plateau, not common, Mar. 1937,
Acocks 4442. Asbestos Mts. Kloof village, 16.2.1812, *Burchell* 2053/2.

B.17. **I. dauensis** *Gillett* sp. nov.

Herba verisimiliter perennis radice sublignoso, ramis patentibus vel repentibus
ad 25 cm. longis, iuventute flavescentibus demum argenteis, pilis densis ad 1·5 mm.
longis appressis, vel saepe subpatentibus. *Folia* ubique strigoso-pilosa, pinnatim
3-foliolata, stipulis sublinearibus, pilosis, 3–8 mm. longis, petiolo *c.*20 mm. longo,
petiolulis 1·2 mm. longis, rhachide ultra foliola lateralia *c.*5 mm. prolongata, foliolis
subaequalibus, vel terminale paulo maiore, ellipticis vel anguste obovatis, apice
rotundatis, mucronatis, ad 24 mm. longis et 12 mm. latis. *Racemi* multiflori,
ubique strigoso-pilosi, ad 10 cm. longi, pedunculo ad 4 cm. longo incluso, bracteis
angustis, mox deciduis, 1·5 mm. longis, pedicellis ad 1 mm. longis, post anthesin
valde reflexis. *Calyx* strigosus, tubo 0·5 mm. longo, dentibus anguste deltoideis,
1·6–1·8 mm. longis. *Vexillum* extus strigosum, diu persistens, cuneato-obovatum,
apice mucronatum, 5 mm. longum, 3·5 mm. latum. *Alae* glabrae brevissime
unguiculatae, ± dolabriformes, 3 mm. longae (immaturae), mox deciduae. *Carina*
ad commissuram strigosa, valde calcarata, haud rostrata, 3·4 mm. longa (immatura),

mox decidua. *Stamina* omnia fertilia, filamentis (maturis) 4 mm, longis, antheris ellipticis, basin muticis, 0·6 mm. longis, mucrone apicale 0·05 mm. longo incluso. *Ovarium* 3 mm. longum, saepius 5-ovulatum, strigosum pilis primo appressis (haud facile visis) deinde patentibus, stylo cylindrico, glabro, paulo ultra mediam fere rectangule curvato, stigmate parvo globoso. *Legumen* rectum, longitudinaliter $\pm$ continue costatum, patente pilosum, valde complanatum, viride stramineum, margine incrassatum, apice acutum, margine inferiore curvato marginem superiorem subrectum ad angulum *c.*40° attingente, *c.*15–18 mm. longum, 2·4 mm. latum, *c.*0·7 mm. crassum. *Semina* 5, rarius 2–4.

Kenya, K1, Yabichu near Ramu on the Daua river, 3°56′N×41°12′E, 360 m., *Acacia- Commiphora* open scrub on pale limestone soils, in a dense growth of annual grasses in an area preserved from grazing for 3 years, corolla dull yellow, 23.5.1952, *Gillett* 13294 K holo EA isotype.

The following may be the same species but has leaves 1–3-jugate, smaller leaflets and flowers, and is without fruits. Wajir 1°45′N, 40°4′E, 240 m., 200 mm. rainfall, red sandy soil, *Commiphora*, etc., scrub with much *Delonix elata*, overgrazed near wells, flowers yellow, 27.5.1952, *Gillett* 13370 A.

This may be conspecific with *I. kelleri* Bak.f., Vierteljahrschr. Nat. Ges. Zur. **49,** 187 (1904); L.T.A. No. 170 which was collected at Abdallah, a tribal area *c.*43°30′E ×6°15′N, some 350 km. N.E. of Ramu. Professor Daniker who kindly loaned several types from the Zurich herbarium was unable to trace the type of *I. kelleri.* However, *I. kelleri*, whose fruits are unknown, is described as "frutex cortice nigrescente" and as having "ovarium glabrum" and if the latter character was correctly observed, can hardly be the same as *I. dauensis.*

B.18. **I. lasiantha** *Desv.*, Ann. Sci. Nat. ser. **1, 9,** 410 (1826).

I. psilostachya Welw. ex Bak., F.T.A. **2,** 84 (1871); L.T.A. No. 187.

Angola, Huilla, Mupande river near lake Ivantala, grassy woods, herb 95–120 cm. tall, lower branches elongated, ascending, flowers, bluish violet, Feb. 1860 *Welwitsch* 2052 (LISU holo) K, BM isotypes of *I. psilostachya*. Benguella plateau, Seculu river valley Kaconda, gregarious in abandoned mandioca plantations, annual herb *c.*1m. tall, flrs., violet-blue, 3.3.1906. *Gossweiler* 4317 BM; Ganda-Caconda, *c.*1700 m., in fields, 50 cm. tall, red-violet flowers, Mar. 1934, *Hundt* 872 BM; Ganda, Alto Catumbela 1350 m., common in bush or forest, 70–100 cm. tall, *Mrs. Faulkner* A287 K (PRE); Membassoco, Cubal, 1050 m., 70–100 cm. tall, *Mrs. Faulkner* A126 K (PRE); coast, *unknown collector* s.n. P holotype.

The fruits of this species have not been seen; they are presumably present on the Lisbon sheet of *Welwitsch* 2052. Baker describes them as linear and *c.*25 mm. long and 2 mm. broad. This species is here placed in subgenus *Amecarpus* principally on account of its close affinity with *I. fanshawei.*

Welwitsch 2027 from Pungo Andongo greatly resembles the present species but its filaments are only 4–5 mm. long.

B.19. **I. fanshawei** *Gillett* sp. nov.

Ab *I. lasiantha* habitu graciliore, ramis cylindricis tenuoribus, stipulis brevioribus, stipellis subnullis, foliolisque minoribus, differt.

Herba ramosa, basin sublignosa, ad *c.*30 cm. alta; rami primo virides demum atro-brunnei, ut folia, racemi, calycesque pilis rigidis medifixis appressis, *c.*0·6–0·8 mm. longis, primo flavescentibus, demum albis vel subfuscis, subdense obsiti. *Folia* 3–7-foliolata; stipulae brunneae, anguste deltoideae vel subulatae, ad 1·5 mm. longae; rhachis ad 2·5 cm. longa, petiolo 3–8 mm. longo incluso, ultra foliola lateralia ad 6 mm. prolongata; stipellae subnullae; petioluli *c.*1 mm. longi; foliola opposita, oblongo-lanceolata, ad 2 cm. longa et 5 mm. lata, apice breviter apiculata, costa mediana superne impressa inferne prominente, pilis ad paginam inferiorem

quam eis paginae superioris crassioribus, rigidioribus, longioribus (ad 1 mm. longis)
et arctius appressis. *Racemi* 10–30-flori, ad 20 cm. longi, pedunculo 4–7 cm. longo
incluso; bracteae subulatae ad 1 mm. longae, mox caducae; pedicelli 1–1·5 mm.
longi ad et post anthesin valde reflexi. *Calyx* tubo inaequale, superne 0·5 mm.,
inferne 1·5 mm. longo, dentibus subulatis, duo superioribus 1·6 mm., inferiore
2·3 mm., aliis 2·0 mm. longis, sinu superiore aliis c.triplo latiore. *Vexillum* mox
caducum extus dense appresse strigosum, pilis medifixis 0·8 mm. longis, ad apicem
emarginatum et breviter apiculatum, late ellipticum, 8 mm. longum, 6 mm. latum.
Alae glabrae, cuneatae, haud unguiculatae, ad 7·5 mm. longae et 2 mm. latae. *Carina*
haud rostrata, infra mediam calcaribus 1 mm. longis praedita, ad margines et versus
apicem strigosa, 8 mm. longa. *Stamina* omnia fertilia, filamentis 6·5–8 mm. longis,
antheris late ellipticis, basin muticis, connectivo fusco, mucrone apicale 0·1 mm.
longo incluso, 0·7 mm. longis. *Ovarium* dense strigosum, c.15-ovulatum, 5 mm.
longum, stylo glabro paulo curvato, 3 mm. longo, stigmate globoso, 0·2 mm.
diametro. [*Legumen* paulo immaturum valde deflexum, c.10-spermum, stramineum,
strigosum, et longitudinale costatum et ad semina umbonatum, apicem versus paulo
curvatum, apice acuto (margine inferiore curvato marginem superiorem rectum ad
angulum c.45° attingente) 30–35 mm. longum 2 mm. latum, c.0·6 mm. crassum.]

S. RHODESIA, Lomagundi district, Trelawney Tobacco Research Station, 1200 m.,
flowers mauve-pink, 27.3.1943 *R. W. Jack* 129 in SRGH 9620 SRGH & 9855, K,
SRGH; 29.3.1944. *Jack* in SRGH 11927 K. N. RHODESIA, Serenje, plateau woodland, thin procumbent woody herb up to 30 cm., flowers pale purple, 18.2.1955,
D. B. Fanshaw 2092 K holo, SRGH isotype. ANGOLA, Bié Province, Domba-
Cuanaval, in open woods amongst *Glumaceae*, a tender annual herblet with bluish
red flowers,1.2.1906, *Gossweiler* 2674 BM; near Munongwe, in open grass-grown
Julbernardia gossweileri woods, Apr. 1906, *Gossweiler* 3365 BM. The fruit is described
from *Jack* in SRGH 9620 SRGH.

B.20. **I. charlieriana** *Schinz*, Verh. Bot. Ver. Brand. **30**, 165 (1888).

α. var. **charlieriana**.

PORT. E. AFR. *Gomes e Sousa* 3669; S. RHODESIA *Jack* 129 SRGH, *Rogers* 6023;
N. RHODESIA *Martin* 601-33, *Gairdner* 210, 434; TROP. BECHUANALAND *Van Son* in
TRV 28914 K, BM; S.W. AFR. *Schinz* 244 isotype, *Rautanen* 446, *Schoenfelder*
S486; EXTRA TROP. BECHUANALAND *Holub* s.n.; N. CENTRAL CAPE, Prieska,
Bryant 1087.

β. var. **scaberrima** (*Schinz*) *Gillett* stat. nov.

I. scaberrima Schinz, Verh. Bot. Ver. Brand. **30**, 163 (1888).

S.W. AFR., *Schinz* 232 isotype; N. CENTRAL CAPE, nr. Kimberley, *M. Wilman* s.n.

γ. var. **sessilis** (*Chiov.*) *Gillett*, Kew Bull. **1955** 573 (1956).

I. linearis DC. var. *sessilis* Chiov., Fl. Somala **2**, 160 (1932).

I. relaxata N.E. Br., Kew Bull. **1925**, 148.

DAHOMEY; SOMALIA; KENYA, K7; TANGANYIKA, T3; ZANZIBAR; PORT. E. AFR.;
ANGOLA; TRANSVAAL.

Except in the Transvaal this variety seems to be confined to coastal sands.

δ. var. **lata** *Gillett* var. nov.

A var. *charlieriana* fructibus brevioribus, latioribus differt. *Herba* annua fere
ubique pilis rigidis medifixis appressis ± sparse obsita, ramis repentibus. *Folia*
1–2-jugata, foliolis anguste oblongis, ad 25 mm. longis. *Racemi* fere sessiles, ad
8 cm. longi, pedicellis post anthesin valde reflexis. *Calycis* dentes c.5 mm. longi,
filamenta aequantes. *Legumen* rectum, valde complanatum, longitudinaliter costatum; costa inter semina haud interrupta, haud centrale (ad suturam inferiorem
crassam quam ad superiorem duplo proximiore); apice subacuto (margine inferiore
curvato marginem superiorem subrectum ad angulum c.60° attingente) 3–5-sper-
mum, 14–18 mm. longum, 2·6 mm. latum, 0·7 mm. crassum.

S.W. Africa, Great Namaqualand, Klein Karas 800 m., flowers "violet", 15.9.1931. *I. Örtendahl* 273 UPS holotype. Damaraland *Een.* s.n. BM.

This plant may eventually, when more material becomes available, be considered a distinct species. Against this *Rogers* 6023, a fragmentary specimen from the Victoria falls, has 4–7-seeded fruits 2·4 mm. wide and thus more or less bridges the gap between var. *lata* and var. *charlieriana. Een.* s.n. BM has until now been identified as *I. senegalensis* on account of its broad short fruits. However, *I. senegalensis* differs in its pedunculate racemes, shorter stamens (3·5 instead of 5 mm.) and in the less appressed indumentum, interrupted lateral ridge, more truncate or rounded apex and much less well developed ventral suture of its fruit.

I. charlieriana is undoubtedly very close to *I. aspera* with which it has been a good deal confused.

B.21. **I. aspera** *Perr. ex DC.*, Prod. **2**, 229 (1825); L.T.A. No. 254.

I. linearis DC., Prod. **2**, 228 fide Hiern, Cat. Welw. Afr. Pl. 209, non *I. linearis* Guill. & Perr.

I. tenella Schum., Beskr. Guin. Pl. 367 (1829).

Senegal, Richard-Tol *Perrottet* (P holo.) K iso, BM iso, (*Bacle* s.n. P holotype of *I. linearis* DC), etc.; Fr. Sudan, *Hagerup* 242 BM; Gold Coast, *Thonning* s.n. C holotype of *I. tenella*; Nigeria, *Barter* s.n., etc.; A.E. Sudan, *Kotschy* 375, *Andrews* 3104, 3124, 3136, etc.

This species has been reported from Angola by Hiern, Cat. Welw. Afr. Pl. **1**. 217, who cites *Welwitsch* 2036, 2036b. A fragment of 2036 without fruits is in BM, it is not possible to be certain, but it appears to be *I. charlieriana* var. *sessilis*.

2 Section **Demissae** sect. nov.

Type species *I. demissa* Taub.

Inflorescentiae parvae, folio breviores; bracteae persistentes; flores parvae; legumina vel reflexa vel patentia.

The three species in this section form a distinctive small group, which to a great extent bridges the gap between subgen. *Amecarpus* and subgen. *Indigofera*. *I. demissa* has very clearly flattened fruits about 2 mm. wide, and, if subgenus *Amecarpus* is to be accepted at all, would clearly appear to belong in it. *I. concinna* has much narrower (*c*.1·3 mm. wide), but still flattened, fruits and, except in fruit, is readily confused with *I. demissa*, while *I. bussei* again must be placed next to *I. concinna*.

B.22. **I. demissa** *Taub.* in Engl. Pflwelt. O. Afr. C 209 (1895); L.T.A. No. 10; F.C.B. **5**, 157.

I. bakeriana Viguier, Not. Syst. **13**, 367 (1949).

B. Congo, Katanga teste Cronquist; Tanganyika, T4, T6, T7; Zanzibar (*Stuhlmann* s.n. B† syntype) *Vaughan* 1933 BM, EA; Port. E. Afr.; Nyasaland, Buchanan 1163 (B† syntype) K, isotype; S. Rhodesia; N. Rhodesia; Madagascar, west, *Perrier* 1417 P, 4691 P, 4864 P, 12530 P, syntypes of *I. bakeriana*.

B.23. **I. concinna** *Bak.*, F.T.A. **2**, 80 (1871); L.T.A. No. 79.

I. mossambicensis Bak.f., J. Bot. **70**, 253 (1932).

Tanganyika, T8, Rovuma river, *Kirk and Meller* s.n. holotype, *Schlieben* 6326 BM; Port. E. Afr., Mozambique *O. Kuntze* s.n., *Dümmer* 4641, *Schlechter* s.n. April. 1895 (B† holo.), BM fragment, isotype of *I. mossambicensis, Pedro & Pedrogao* 3116.

B.24. **I. bussei** *Gillett* sp. nov.

Ab *I. concinna* Bak. partibus omnibus multo majoribus differt.
Herba ramosa caulibus diffusis, sparse appresse pilosis, pilis albidis *c.*0·2 mm.
longis. *Folia* 3–5-jugata, ubique appresse pilosa, rhachide ultra foliola lateralia
opposita prolongata, stipulis brunneis subulatis, *c.*3 mm. longis, basin ad petiolum
1 mm. connatis, petiolo 2–6 mm. longo, stipellis brunneis filiformibus ad 1·5 mm.
longis, petiolulis *c.*0·5 mm. longis, foliolis mucronatis sursum accrescentibus,
basalibus minimis ± ellipticis, *c.*4 mm. longis, 2·5 mm. latis, terminalibus maximis,
oblanceolatis, *c.*14 mm. longis, 4·5 mm. latis. *Racemi* 6–10-flori, foliis breviores,
15–20 mm. longi, pedunculo 5–8 mm. longo incluso, bracteis filiformibus *c.*2 mm.
longis, persistentibus, pedicellis *c.*2 mm. longis, post anthesin paulo, haud valide,
reflexis. *Calyx*, tubo pallide fusco, *c.*0·5 mm. longo, nervis 5 atro-brunneis in
dentes angustos 3–3·5 mm. longos euntibus. *Vexillum* extus pubescens, late ovatum
apice subacuminatum, 4 mm. longum, 2·8 mm. latum. *Alae* rotundato triangulariae,
breviter unguiculatae. *Carina* subtriangularia, margine pubescens, calcarata, haud
rostrata. *Stamina* omnia fertilia, *c.*4·5 mm. longa, antheris 0·7 mm. longis, mucrone
apicale brunneo 0·15 mm. longo et mucrone basale brunneo 0·03 mm. longo
inclusis. *Stylus* glaber subcomplanatus, semicirculare curvatus, stigmate capitato.
Legumen rectum, breviter subsparse strigosum, 5·9-spermum, 12–14 mm. longum,
*c.*1·4 mm. latum, *c.*0·6 mm. crassum.

TANGANYIKA, T8, probably near Lindi, 1903, *W. Busse* 2361 EA holotype.

Subgenus C. Indigofera: Key to sections and subsections

Stipules very large, approaching the size of a leaflet, flowers single or two together
 in the axils of leafy bracts and often hidden by their stipules, or in short 1–3-
 flowered racemes. Fruits 2–4-seeded . . . . 1. **Latestipulatae.**

Stipules much smaller than a leaflet in size, or, if enlarged, inflorescence not as
 above:
 Inflorescence a lax or condensed panicle, or flowers single in the leaf axils; pod
  usually rather short and 1–4-seeded:
  Pedicels less than 4 mm. long . . . . 2. a. **Paniculatae.**
  Pedicel filiform, over 4 mm. long . . . . 2. b. **Trichopodae**

Inflorescence an axillary raceme:
 Leaves compound, or, if simple, then fruiting pedicels not reflexed:
  Leaflets opposite:
   Racemes borne in the axils of normal leaves in the upper part of the plant:
    Fruiting pedicels erect or spreading, or a few of the lower ones only
    reflexed:
    Petiole elongated, persisting after the leaflets have fallen . . . .
                  3.a. **Juncifoliae.**

    Petiole not elongated, not persisting after leaflets have fallen:
     Axis of inflorescence a spine . . . . 3.e. **Spinosae.**

    Axis of inflorescence not a spine:
     Bracts persistent; peduncle under 1 cm., usually much less;
      calyx deeply divided, about as long as the stamens which
      are up to 4 mm. long; fruits deflexed suberect or spreading;
      leaflets not over 14 mm. long . . . . B.2. **Demissae.**

    Bracts not persistent or if so, other characters not as above:
     Peduncle short, less than half as long as a fruit, or, if fruit
      very short, under 4 mm.:

Fruits erect, at least when mature, usually more than 3 to an inflorescence . . . . 3.b. **Brevi-erectae.**

Fruits erect, spreading or reflexed rarely more than 3 to an inflorescence . . . . 3.f. **Brevipatentes.**

Peduncle longer, as long as a fruit, or over 4 mm.:

Whole plant including standard glabrous or almost so, pod 1-seeded flattened, inflorescence 4–7-flowered . . . . . 3.c. **Anomalae.**

Standard, at least, not glabrous:

Fruiting pedicels and fruits ± erect; fruiting peduncles rigid, often glabrescent, rather wiry, inflorescence rather open, usually with fewer than 12 flowers; glandular hairs absent:

Calyx not more than half as long as stamens, stipules linear-subulate, 1-veined, leaflets often more than 5, usually small (i.e. under 50 sq. mm.) hairs often short and appressed . . . . . 3.d. **Dissitiflorae.**

Calyx more than half as long as stamens; stipules lanceolate, more than 1-nerved, leaflets 1–5, often more than 50 sq. mm., hairs long and spreading, not dark brown or black . . . . . 3.g. **Pilosae.**

Fruiting pedicels and fruits more or less spreading; inflorescence lax or dense:

Fruit not glabrous or, if nearly so, fruits 2-seeded and conspicuous dark hairs in leaflets axils:

Medifixed hairs all white; no dark tufts in leaflet axils:

Glandular multicellular hairs present . . . . 3.h. **Viscosae.**

Glandular multicellular hairs absent . . . . 3.i. **Centrae.**

Some medifixed hairs dark brown or black, or conspicuous dark tufts in leaflet axils, or both . . . 3.j. **Atratae.**

Fruit glabrous, more than 2-seeded . . . . 3.k. **Psiloceratiae.**

Fruiting pedicels reflexed, inflorescence more than 6-flowered, usually more or less dense:

Fruit and ovary glabrous or almost so . . . . . . 3.k. **Psiloceratiae.**

Fruit pubescent, at least when young:

Leaves not gland-dotted below:

Calyx lobes not more than twice as long as tube; corolla much longer than calyx; indumentum on standard closely appressed and parallel, giving a shiny appearance; pods often curved . . . . 3.m. **Tinctoriae.**

Calyx lobes many times as long as tube; corolla not twice as long as calyx: indumentum of standard not dense close-appressed and shiny; pods 3–8-seeded, straight . . . . 3.n. **Hirsutae.**

Leaves gland-dotted below . . . . 3.o. **Microcarpae.**

Racemes borne in the axils of usually reduced leaves on the basal parts of
the shoot, below the normal foliage leaves, usually appearing
precociously after fires . . . . 3.l. **Geanthae.**

Leaflets alternate . . . . 3.p. **Alternifoliolae.**

Leaves simple, fruiting pedicels reflexed . . . . . . 3.q. **Simplices-reflexae.**

Section C.1. **Latestipulatae** (*Bak.f.*) *Gillett.*

"Group" *Latestipulatae* Bak.f. L.T.A. 112 (1926) emend.

" Group" *Opertiflorae* Bak.f.1.c.

Type species *I. berhautiana.*

Centre of distribution S.E. Tropical Africa.

Indumentum including long, delicate, soft, subsimple hairs, as well as stiffer
medifixed hairs. Leaves 1–7-foliolate, if compound, pinnate; the rhachis prolonged
beyond the opposite lateral leaflets. Stipules foliaceous, large, approaching the
leaflets in size. Flowers 1–3 in the axils of leafy bracts. Calyx divided. Fruit short,
usually ± erect, sometimes ± flattened, 2–5 (usually 2)-seeded, the endocarp
occasionally (in sp. 8) with dark spots but usually having two dark brown stripes on
each valve. Seeds closely pressed together.

This section approaches §§**Pilosae** on the one hand and §§**Dissitiflorae** on the
other. It is not known to occur outside Africa. Enlarged stipules are also found in a
group of S. African species, of which *I. dimidiata* reaches Nyasaland. These
however have long racemes with deflexed fruits and digitately 3-foliolate leaves.
Peduncles filiform, many times as long as calyx; indumentum rather sparse:

Leaves simple . . . . 1. *ischnoclada.*

Leaves pinnately 3-foliolate . . . . 2. *gairdnerae.*

Peduncles less than 3 times as long as calyx: indumentum copious:

Leaves (bracts) subtending the flowers, or groups of 2–3 flowers, not usually
compacted to form a strobilus, having a well developed rhachis and lamina,
the stipules free almost to the base:

Flowers 2–3 together in the axil of the subtending leafy bract:

Foliage leaves 3–7-, usually 5-, foliolate:

Calyx lobes gland-tipped;bracts subtending pairs of flowers passing
imperceptibly into the ordinary foliage leaves:

Calyx glands *c.*0·2 mm. in diameter, similar glands at the tips of most
leaflets and stipules; lower pair of lateral leaflets often much
reduced, ± subulate . . . . 3. *arenophila.*

Calyx glands *c.*0.1. mm. in diameter, or less, most leaflets and stipules,
apart from some next to the flowers, without gland tips; lower
lateral leaflets not much reduced . . . . 4. *kuntzei.*

Calyx lobes not gland-tipped; inflorescences aggregated into dense ±
capitate axillary panicles in which the bracts have each one pair of
much reduced lateral leaflets and are very distinct from foliage leaves
5. *faulknerae.*

Foliage leaves 3-foliolate, similar to, or passing gradually into, the bracts:

Stipules over twice as long as petiole, at least in the compound leaves:

Leaflets ovate-oblong, not attenuate at base; stipules broadly ovate-
lanceolate 1·5–3 times as long as wide . . . . 6. *eylesiana.*

Leaflets ± oblanceolate, attenuate at base; stipules lanceolate, 4–5 times
as long as wide . . . . 7. *burttii.*

Stipules less than twice as long as petiole, *c.*4 times as long as wide; leaflets ± oblanceolate, attenuate at the base . . . 8. *inhambanensis.*

Flowers single in the axil of the subtending leafy bract:

Leaves subtending inflorescence branches, and most of the others, 1-foliolate; stipules 5 or more times as long as the up to 2 mm. long petiole
9. *taborensis.*

Leaves subtending inflorescence branches usually 3-foliolate, lower leaves 3–7 foliolate, stipules 2–3 times as long as petiole . . . . 10. *berhautiana.*

Leaves (bracts) subtending the flowers aggregated into a strobilus, each reduced to a pair of large stipules which are united, usually to above the middle; with or without a rudimentary rhachis . . . . 11. *strobilifera.*

Foliage leaves 3–5-, very rarely 7-foliolate, the terminal leaflet larger than the others; strobili pedunculate often lengthening to over 3 cm. in fruit . . . .
α. ssp. *strobilifera.*

Foliage leaves 5–9-foliolate, the terminal leaflet hardly larger than the others; strobili sessile or shortly pedunculate, not, or very rarely, lengthening to over 3 cm. in fruit . . . . β. ssp. *lanuginosa.*

C.1. 1. **I. ischnoclada** *Harms,* Not. Bot. Gart. Berl. **11,** 812 (1933).

Tanganyika, T6, Mahenge district, near junction of Kilombero and Luwegu with the Rufiji river, *c.*400 m., sandy dry forest, 13.6.1932, *Schlieben* 2370 (B† holo.) BM isotype. T8, 140 km. W. of Lindi, 400 m., *Schlieben* 6327 BM. Nyasaland, Nselema village Zomba, weed in old gardens, 23.4.1955, *E. A. Banda* 86.

C.1. 2. **I. gairdnerae** *Hutch. ex Bak. f.,* L.T.A. 115 (1926) No. 73.

A full description follows, as that given by E. G. Baker is totally inadequate.

Species propter pedunculos longos, tenellos, pilos sparsos longos patentes et stipulas magnas latasque insignis. Ab *I. ischnoclada* Harms foliis pinnatim 3-foliolatis differt.

Herba annua ad 20 cm. alta, ramis tenuibus flavis patentibus, pilis sparsis patentibus ad 1·5 mm. longis conspersis. *Folia* pinnatim trifoliolata; stipulae oblongo-lanceolatae, ad 6 mm. longae, 2 mm. latae, pilis sparsis marginalibus longis exceptis, glabrae, apice, primo glanduloso-apiculatae, acutae; petiolus flavus, sparse longeque pilosus, ad 5 mm. longus; rhachis ultra foliola lateralia ad 3 mm. prolongata; foliola elliptica vel obovata, terminale ad 9 mm. longum, 6 mm. latum, lateralia ad 7 mm. longa, 4 mm. lata. omnia untrinque pilis longis basifixis sparsis conspersa, superne etiam pilis medifixis appressis sparse vestita, apice, primo glanduloso apiculata, obusa. *Racemi* axillares parcissime pilosi 1–3-flori, pedunculo tenello ad 22 mm. longo, bracteis minutis mox deciduis, pedicellis *c.*1 mm. longis. *Calyx* sparse longeque pilosus 1·9 mm. longus, tubo 0·8 mm. longo incluso, dentibus anguste deltoideo-acuminatis, apice primo glandulosis. *Corolla* extus puberula. *Filamenta* ad 3 mm. longa. *Legumen* immaturum ± erectum, glabrum ± complanatum, ad 8 mm. longum, 1 mm. latum, 2–4-spermum.

N. Rhodesia, Sesheke, 24°45′E × 17°30′S, found only on an island in the flooded Zambesi, April, flower pink, *Miss A. E. Gairdner* 510 holotype.

These two species form a transition between this section and subsection *Pilosae.* They are closely related and may prove to be conspecific when more material is available. Besides its simple leaves *I. ischnoclada* has a scantier indumentum and rather shorter peduncles and pods than *I. gairdnerae.*

C.1. 3. **I. arenophila** *Schinz,* Bull. Herb. Boiss. **6,** 524 (July 1898); L.T.A. No. 81.

I. antennulifera L. Bolus, Ann. Bol. Herb. **1,** 188 (1915).

I. sericea Benth. ex Bak., F.T.A. **2,** 76 (1871) pro min. parte, quoad plantam afr. austr.; non *I. sericea* L.

"*I. polycarpa Benth. ex Harv.*" sensu R. E. Fries, Schwed. Rhod. Kong. Exp. 77 et L.T.A. No. 87, non Benth. ex Harv., nec Willd.

S. Rhodesia, nr. Victoria falls, (*Flanagan* 3089 BOL holotype of *I. antennulifera*), *Rogers* 131600, *Wilman* 16704, *Eyles* 1132, *R. E. Fries* 22 UPS 22a UPS; N. Rhodesia, S. Prov. *Fanshawe* 285, Barotseland *Gairdner* 523, *Codd* 7562, *Ag. Dept.* 12; S.W. Afr., *Rautanen* 215 isotype, *Chapman and Baines* s.n., *Boss* in TRV 35993, *Dinter* 7275/A BM.

C.1. 4. **I. kuntzei** *Harms* in O. Kuntze Rev. Gen. 3: 2, 51 (18.9.1898).

Tanganyika, 17·5 Km. W. of Lindi, *Milne-Redhead and Taylor* 7647; Port. E. Africa, Mozambique, 5.4.1894, *O. Kuntze* s.n. K. isotype.

C.1. 5. **I. faulknerae** *Gillett* sp. nov.

Species *I. strobiliferam* et *I. arenophilam* accedens. Ab illa bracteis inflorescentiae laminiferis et stipulis bractearum fere ad basin liberis, ab ea inflorescentia congesta et foliolis, stipulis calycis lobisque haud glandulosis differt.

Herba annua, ramis pilosis viridibus saepius decumbentibus ad 25 cm. longis. *Folia* 3- vel saepius 5-foliolata, ubique, sed haud dense, sericeo pilosa; stipulae inaequales, versus petiolum margine paulo curvato vel fere recto, externe basin late cordatae, apice acuminatae, ad 9 mm. longae, 3–4 mm. latae; petiolus 2–3 mm. longus; rhachis ultra foliola lateralia opposita *c.*5 mm. prolongata; foliola late lanceolata, superne, et pilis subsimplicibus sericeis sparsis, et pilis medifixis rigidis appressis, obsita, terminale, lateralibus paulo majus, ad 15 mm. longum, 7 mm. latum. *Inflorescentiae* paniculae sericeo-pilosae densae, foliosae, subsessiles, axillares; flores binati in axillis bractearum foliosarum, pedunculo 1 mm. longo, pedicellis duobus 2 mm. longis, post anthesin reflexis, complanatis, superne incrassatis; bracteae, stipulis illis foliorum normalium subsimilibus, sed basin connatis, rhachidem superantibus, foliolo terminale lanceolato stipulis $\pm$ aequilongo, foliolis lateralibus, unijugis, valde reductis, filiformibus, *c.*2 mm. longis. *Calyx* sericeo-pilosus, tubo *c.*0·5 mm. longo, lobis anguste lanceolatis inaequilongis, ventrale 3 mm. longo, aliis 2 mm. longis. *Vexillum* obovatum, extus minute pubescens, 3 mm. longum, 2 mm. latum. *Alae* glabrae. *Carina* ad marginem pubescens, calcarata, haud rostrata. *Stamina* omnia fertilia, filamentis *c.*3·5 mm. longis, antheris basin muticis, *c.*0·4 mm. longis, mucrone apicale fusco *c.*0·08 mm. longo incluso. *Ovarium* sericeum *c.*1 mm. longum, 3–4-ovulatum, stylo in parte terminale curvato. *Legumen* paulo curvatum, dense breviter lanuginosum, *c.*6 mm. longum, 1·3 mm. latum, 3–4-spermum, acutum, vix compressum.

Port. E. Afr., Mocuba-Muobede Rd., *c.*36°45′E × 16°35′S, *c.*300 m., on edge of forest among grass, spreading and decumbent, flowers bright crimson, 6.4.1948, *Mrs. H. G. Faulkner* Kew No. 227, holotype; Manica e Sofala, Belas Chimoio, annual herb of stony slopes, 2.3.1948, *Garcia in Mendonça* 467 LISC. Nyasaland, Shiri highlands, *Buchanan* 335.

This species is named after Mrs. H. G. Faulkner whose excellent collections in Angola, Portuguese East Africa and Tanganyika have greatly enriched the Kew and other herbaria.

C.1. 6. **I. eylesiana** *Gillett* nom. et stat. nov.

I. sericea Benth. ex Bak. forma *australis* Bak.f., J. Bot. **41,** 235 (1903); L.T.A. No. 62 forma.

Prostrate, up to 1 m. in diameter. The stipules are up to 10 mm. long and 6 mm. wide. The calyx lobes are all of the same length. Fruit straight, pubescent, *c.*5 mm. long, 2 mm. wide, 3–4-seeded.

S. Rhodesia, Lomagundi distr. *Archdale in Eyles* 5169, *Agr. Dept.* 1146; Salisbury distr. *Eyles* 925, *Gilliland* 131, K, BM, *Whellan* 634; Marandellas distr. *Corby* 138 K, SRGH, *Rattray* in SRGH 4450 BM; Umtali distr. *Teague* 490; Bulawayo distr. *Rand* 59 BM holotype; Victoria distr. *Monro* 1289 BM; district not known *Hislop* Z83; N. Rhodesia, S. Prov. nr. Choma, *Robinson* 1182.

Several of the collectors note that this species grows on sand.

C.1. 7. **I. burttii** *Bak.f.*, J. Bot. **70**, 251 (1932).

Tanganyika, T4, Tabora distr., *Peter* 34847, 34848; T5, Manyoni distr., *B.D. Burtt* 3434 BM, holo. K, isotype, 3580, 5063; T7, Iringa distr., *Greenway* 3389.

This species varies a good deal and requires further investigation.

C.1. 8. **I. inhambanensis** *Klotzsch*, Pet. Moss. Bot. 48 (1861); L.T.A. No. 86.

I. polycarpa Benth. ex Harv., Fl. Cap. **2**, 191 (1862), non Willd. ex Spreng. (1826).

Port. E. Afr., Inhambane, (*Peters* s.n. B† holotype); Delagoa bay, *Forbes* s.n. holotype of *I. polycarpa* Benth. ex Harv., *Gomes e Sousa* 159A, *Schlechter* 11560; S. Rhodesia, Wankie, *Levy* 47 PRE, 176 K; Transvaal, Nelspruit distr., c.600 m., *Codd & De Winter* 4922. *Liebenberg* 3289; Natal, Inanda, *Wood* 1296, *Gerrard* 509.

The type of *I. inhambanensis* having been destroyed, further collections at Inhambane are desirable to remove any traces of doubt in this identification. Young leaflets and calyx lobes are sometimes slightly gland-tipped, showing an approach to *I. arenophila*.

C.1. 9. **I. taborensis** *Gillett* sp. nov.

Species *I. berhautianae* similis; ut in clavi distinguitur.

Suffrutex sublignosus ramosissimus, in sicco nigrescens, ramis brunneis cylindricis pilosis. *Folia* simplicia, vel inferiora rarius pinnatim trifoliolata; stipulae magnitudine foliolis aequales, subrhomboideae, basin inaequales, apice acuminatae, subtus et ad margines dense sericeo pilosae, superne fere glabrae, ad 10 mm. longae, 4 mm. latae; petiolus 1–2 mm. longus; lamina late lanceolata, ad 10 mm. longa, 4 mm. lata, utrinque sericeo pilosa, superne etiam pilis perpaucis, rigidis, medifixis, appressis, praedita. In foliis 3-foliolatis, rhachis ultra foliola lateralia 0·5–1·0 mm. prolongata; foliola subaequimagna. *Flores* in axillis foliorum solitarii, 1 mm. pedicellati. *Calyx* dense sericeo pilosus, c.5 mm. longus, tubo 0·5 mm. longo incluso, lobis lanceolatis, apice longe acuminatis, 1 mm. latis. *Vexillum* glabrum, ovale, basin paulo attenuatum, apice obtusum, c.4 mm. longum 2·5 mm. latum. *Alae* glabrae, breviter unguiculatae, basin auriculatae, suboblongae, c.3 mm. longae, 0·7 mm. latae. *Carina* fere glabra (margine parcissime strigosa), haud rostrata, 0·5 mm. calcarata, 4 mm. longa. *Stamina* omnia fertilia, c.4 mm. longa, antheris basin muticis, suborbicularibus c.0·5 mm. longis, mucrone apicale fusco c.0·1 mm. longo incluso. *Ovarium* 1 mm. longum, glabrum, 2-ovulatum, stylo, base pauce piloso excepto, glabro, 3·5 mm. longo, 1 mm. infra stigma rectangulare curvato, stigmate capitate 0·3 mm. lato. *Legumen* haud visum.

Tanganyika, T4, Tabora distr., 72 km. N. of Tabora, shrub, medicine from roots used for chest trouble, 8.12.1926, *G. B. Wallace* 8. S. of Kombe, c.31°30′E, 5°5′S, 1100 m., subshrub, flowers light red, 24.1.1926, *A. Peter* 35373, K, holo. (B, isotype). Between Kombe and Usinga, km. 1003·6 (from Dar-es-Salaam) on the railway, 1100 m., 26.1.1926, *A. Peter* 35599.

C.1. 10. **I. berhautiana** *Gillett*, Kew Bull **1955**, 573 (1956).

I. sericea Benth. ex Bak., F.T.A. **2**, 76 (1871) pro maj. parte, quoad pl. guineenses; L.T.A. No. 62 pp.; non *I. sericea* L. (1771).

"I. aff. strobilifera Hochst." sensu Chevalier Expl. Bot. Afr. Occ. Fr. 177.
SENEGAL; GOLD COAST; FR. NIGER COL.; NIGERIA.

C.1. 11. **I. strobilifera** *(Hochst.) Hochst. ex Bak.*, F.T.A. **2,** 75 (1871); L.T.A.
No. 64.

α. ssp. **strobilifera.**
Eilemanthus strobilifer Hochst., Flora **29,** 593 (1846).
I. lupulina Bak., Kew Bull. **1897,** 254; L.T.A. No. 63.
FR. SUDAN, *Davey* 122; FR. NIGER COL., *Hagerup* 475 BM p.p.; NIGERIA, *Dalziel*
330; CAMEROONS, *Talbot* s.n. BM; A.-E. SUDAN, *Pfund* 79, 96, 389, *Kotschy* (1838)
236 K, BM isotypes; TANGANYIKA, T1 *Koritschoner* 2210 EA, T4 *Lindeman* 351
BM, EA, 636 EA, *Smith* 1126, T5 *Burtt* 3651, *Lambert* 6; NYASALAND, *Whyte* 336
holotype of *I. lupulina, Benson* 1202, Exell, Mendonca & Wild 1111; S. RHODESIA,
Wankie *Levy* 6; N. RHODESIA, Livingstone *Rogers* 7026, 13160, Sesheke *Gairdner*
464. An inland plant.

β. ssp. **lanuginosa** *(Taub. ex Bak.f.) Gillett* stat. nov.
I. lanuginosa Taub. ex Bak.f., J. Bot. **41,** 236 (1903); L.T.A. No. 65.
KENYA, K7 *Hildebrandt* 927b, etc.; TANGANYIKA, T3 *Holst* 2066 isotype, T6,
T8 *Braun* 1336 EA; ZANZIBAR; PEMBA; PORT. E. AFRICA, Mozambique *Forbes* s.n.
Predominantly a coastal plant. Both subspecies are usually (or always ?) found on
sandy soils.

Section C.2. **Paniculatae** *(Bak.) Gillett.*
a. subsection **Paniculatae.**

"Group" *Paniculatae* Bak., F.T.A. **2,** 67 (1871) p.p., emend; L.T.A. 110.
"Group" *Terminales* Bak.f., L.T.A. 97.
"Group" *Capitatae* Bak.f., L.T.A. 97.
Lectotype species *I. pulchra* Willd.
Centre of distribution West and West Central Tropical Africa.
Leaves various, the stipules not larger than normal not approaching a leaflet in
size. Inflorescence a panicle, sometimes condensed and subcapituliform, the bracts
passing more or less gradually into the foliage leaves; hairs on inflorescence and
calyx often brown; pedicels not above 4 mm. long, often jointed and bracteolate at
the joint as though the lower part may represent the peduncle of a raceme reduced
to a single flower. Calyx deeply divided. Fruits usually short (1–4-seeded) and
broad (except in *I. trialata*), erect or spreading; not reflexed, straight. Endocarp
occasionally diffusely pigmented but not definitely spotted. Seeds, where more than
one, closely pressed together.
Besides the species treated here the section includes *I. mysorensis* Rottb. from
peninsular India.
More or less paniculate inflorescences may also be found in §**Latestipulatae,**
and §§**Dissitiflorae.**

Leaves always simple, the lower more than 2 cm. long:
    Fruiting calyx 4–5 mm. long, with conspicuous rufous hairs . . . . 1. *polysphaera.*
    Fruiting calyx 1–2 mm. long:
        Fruit *c.* 5 times as long as wide, stem 3-winged . . . . .2. *trialata.*
        Fruit up to 3 times as long as wide, stem ridged, not winged . . . 3. *paniculata.*

Inflorescence lax, its smaller divisions often longer than the subtending bract; hairs on upper leaf surface at an angle of *c*.30° with midrib (or sometimes absent); stamens *c*.3 mm. long . . . . . . .α. ssp. *paniculata*.

Inflorescence condensed, its smaller divisions shorter than subtending bract; hairs on upper leaf surface at *c*.60° to midrib; stamens *c*.4 mm. long . . . . β. ssp. *gazensis*.

Lower leaves often pinnate, simple leaves, if any, usually under 2 cm. long:

Fruiting calyx lobes much broadened, wider than the usually 1-seeded fruit, which is hidden within them:

Straggling herb; fruiting inflorescence compact, usually over 15 mm. wide and less than twice as long; without glands or with glands less frequent than the normal hairs . . . . 4. *macrocalyx*.

Stiff semiwoody plant: fruiting inflorescence less compact, usually under 15 mm. wide and often more than twice as long; with numerous short glands between the less frequent normal hairs . . . . 5. *terminalis*.

Fruiting calyx lobes not so wide as fruit:

Inflorescences all compact, subglobose, congested panicles; no separate flowers away from the others in leaf axils:

Inflorescences at the ends of shoots which are longer than the inflorescence is wide, fruiting calyx *c*.6 mm. long, or more:

Leaves (the lower ones not seen) simple, or, less often 3-foliolate, leaflets 3–5 m. long; inflorescences 3–4 cm. across; fruiting calyx *c*.11 mm. long . . . . 6. *megacephala*.

Leaves 5–9-foliolate, occasionally a few 3-foliolate, leaflets rarely over 2 cm. long; inflorescence under 3 cm. across; fruiting calyx *c*.6–8 mm. long:

Hairs on stems, leaf rhachises and leaflet midribs numerous short, closely appressed, those on calyx usually numerous, long silky, spreading, parallel; stems and leaf rhachises ± cylindrical; pod silky, usually 1-seeded, *c*.2·5 mm. long . . . . 7. *capitata*.

Hairs on stems, leaf rhachises and leaflet midribs sparse, ± spreading, those on calyx sparser than in *I. capitata* and irregular; upper stems and leaf rhachises markedly angled; pod glabrous usually 2-seeded, *c*. 3–4 mm. long . . . . 8. *paracapitata*.

Inflorescences mostly lateral, sessile in leaf axils or on side shoots which are shorter than the inflorescence is wide; fruiting calyx *c*.3 mm. long:

Leaflets up to 3-jugate:

Calyx longer than pod; some hairs on inflorescence usually brown; erect bushy herb . . . . 9. *congesta*.

Calyx slightly shorter than pod; hairs on inflorescence all white; herbaceous shoots from a woody rootstock . . . . . 10. *lotononoides*.

Leaflets 5–6-jugate . . . . 11. *dasycephala*.

Inflorescences much less compact, many flowers single in the axils of leaf-like bracts (or simple leaves):

Fruiting calyx 3–4 mm. long, ± as long as fruit; indumentum of calyx brownish:

Indumentum on fruit dense, golden brown, ± appressed; fruiting calyx segments up to *c*.1 mm. broad:

Leaves all pinnate, the longer ones *c*. 17-foliolate . . . . . 12. *brassii*.

Leaves often simple or 3-foliolate, longer ones *c*.9-foliolate . . . .
13. *pulchra*.

Fruit glabrous; fruiting calyx segments *c.*1·5 mm. or more broad . . . . .
14. *vanderystii.*

Fruiting calyx not above 2·5 mm. long:
Lobes of fruiting calyx subequal, not more than half as long as fruit:
Fruit 2–4-seeded, most lower leaves pinnate (up to 11-foliolate):
Stiffly erect annual; leaflets 4 or more times as long as wide; branches of inflorescences leafless and flowerless below, crowded with simple or trifoliolate leaves or bracts and flowers in their upper half; hairs on calyx and fruit dense, *c.*0·5 mm. long, often black or dark brown . . . . 15. *nigricans.*

Woody suffrutex, leaflets *c.* twice as long as wide, branches of inflorescence without a leafless lower portion; hairs on calyx and fruit sparser, *c.*0·2 mm. long, white . . . . . . 16. *oubanguiensis.*

Fruit 1-seeded ± oval; leaves simple; indumentum sparse; herbaceous shoots from a woody rootstock . . . . 17. *leptoclada.*

Two upper lobes of fruiting calyx much wider than the others, about ¾ as long as fruit; hairs on calyx and fruit white . . . . 18. *bracteolata.*

C.2.a.1. **I. polysphaera** *Bak.*, Kew Bull. **1895,** 65; L.T.A. No. 23; F.C.B. **5,** 128.

I. dewevrei Mich., Compt. Rend. Soc. Bot. Belg. **36,** 54 (1897).

I. baoulensis Chev., Bull. Soc. Bot. Fr. **58** mem. **8,** 158 (1912).

"*I. trachyphylla Benth.*" sensu Chipp, List Herb. Pl. & Undershrubs of Gold Coast, etc.; non Benth. ex Oliv.

Ivory Coast, *Chevalier* 22146 isotype of *I. baoulensis*; Gold Coast; Nigeria; Fr. Cameroons; Oubangui Chari; Fr. Congo; Belg. Congo, *Dewevre* 197 BR holotype of *I. dewevrei*; A.E. Sudan; Uganda, U1, U3; Tanganyika, T4, T7; N. Rhodesia, *Carson* 70 holotype; Angola.

Ujor in FHI. 31607 from Share forest reserve Ilorin Nigeria has the inflorescence much more condensed than usual, and may represent a distinct variety.

C.2.a.2. **I. trialata** *Chev.*, Bull. Soc. Bot. Fr. **58** mem. **8,** 157 (1912); L.T.A. No. 32.

Ivory Coast, *Chevalier* 22173 isotype; Gold Coast.

C.2.a.3. **I. paniculata** *Vahl ex Pers.*, Syn. **2,** 325 (1807); F.C.B. **5,** 126 t.9; non sensu D.C., Prod. **2,** 222; nec Bak., F.T.A. **2,** 71; nec L.T.A. No. 20. See Tisserant, Bull. Mus. Hist. Nat. Par. 2nd ser. **2,** 680 (1930).

α. ssp. **paniculata.**

I. *procera* Schum. et Thonn., Beskr. Guin. Pl. 365 (1829); L.T.A. No. 21.

Gambia; Port Guinea; Fr. Guinea; Sierra Leone; Ivory Coast, *Chevalier* 21889P; Gold Coast (? *Thonning* s.n. P holotype) (*Thonning* s.n. C holotype of *I. procera*); Togoland; Dahomey; Nigeria; Br. Cameroons; Fr. Eq. Afr.; B. Congo; Oubangui Chari; Uganda, U2–4; Kenya, K7; Tanganyika, T1, T6; Zanzibar; Port. E. Afr., Zambezia, *Mendonça* 2074 LISC; Angola.

β. ssp. **gazensis** (*Bak.f.*) *Gillett* comb. nov.

I. *procera* Schum. et Thonn. var. *gazensis* Bak.f., L.T.A. 103 (1926).

Tanganyika, T6 *Drummond & Hemsley* 1489, *Rounce* 526EA. T7 *Stolz* 271, *St. Clair Thompson* 1008, *Davies* 579, 632, *Hornby* 14; Port. E. Afr., Gazaland, Cuipetzoma *Swynnerton* 1485 K iso. BM holotype, nr. Nampula *Torre* 1284 LISC, Manica, Mavita *Barbosa in Mendonça* 1549 LISC, 1658, LISC, *Torre* 4386; Nyasaland, *Buchanan* 324, 481, 730, *Faulkner* 397, *Adamson* 92; S. Rhodesia, *Johnson* 182, *Martineau* 591 SRGH, *Hack* 130/50 SRGH.

Tisserant, who examined Persoon's type of *I. paniculata* in herb. Jussieu in Paris, states that it is completely identical with *I. procera*. Apparently De Candolle when he took up the species in the Prodromus added a description of a fruiting specimen of some other, unidentified, species, this error being repeated by Baker, E. G. Baker and others.

C.2.a.4. **I. macrocalyx** *Guill. et Perr.*, Fl. Seneg. 175, t.46 (1832); L.T.A. No. 11.

"*I. lotononoides Bak.f.*" sensu Chevalier, Expl. Bot. Afr. Occ. Fr. 174 non Bak.f.

Senegal *Perrottet* 193 BM, Bakel (*Leprieur* s.n. P. holotype); Port Guinea; Sierra Leone; Ivory Coast, *Chevalier* 21423 P; Fr. Sudan; Nigeria; Fr. Chad. Terr.

C.2.a.5. **I. terminalis** *Bak.*, F.T.A. **2**, 70 (1871); L.T.A. No. 12, including var. *chevalieri* Bak.f., L.T.A. 99 (1926).

Senegal, *Heudelot* 365 holotype; Fr. Sudan, *Chevalier* 187 holotype of var. *chevalieri*, *Roberty* 3467 IFAN; Ivory Coast, *Aubreville* 1816 P.

Although the types of *I. terminalis* appear very distinct from the Nigerian plants of *I. macrocalyx* various intermediates occur in the area where both species are found. For instance, *Esperito Santo* 1447, 2819 are specimens of *I. macrocalyx* from Port. Guinea which have the glandular hairs characteristic of *I. terminalis*. Apart from this *I. macrocalyx* seems in any case to be more variable in the western part of its range.

C.2.a.6. **I. megacephala** *Gillett*, Kew Bull. **1955**, 574 (1956).

Fr. Guinea.

C.2.a.7. **I. capitata** *Kotschy*, Sitzb. Ak. Wien. Math. Nat. **51**, Abt. 2, 365 t. 6A ("Plantae Binderianae") (1865); F.T.A. **2**, 75 pro maj. parte; F.W.T.A. ed 1. **1**, 393; Berhaut, Fl. Seneg., 51; non sensu Cronquist, F.C.B. **5**, 152, t.11; nec R. Graham in Wall. Cat. 5490 (nomen).

Senegal, *Etesse* s.n. P; Port. Guinea; Fr. Guinea; Sierra Leone; Fr. Sudan; Nigeria; Oubangui Chari; A.-E. Sudan (*Binder* 56, 70 W syntypes); Uganda U1, 3; N. Rhodesia, Abercorn distr.; Angola, Malange distr., *Gossweiler* 1357 K, BM.

C.2.a.8. **I. paracapitata** *Gillett* sp. nov.

"*I. capitata Kotschy*" sensu Bak., F.T.A. **2**, 75 pro minore parte, sensu Cronquist, F.C.B. **5**, 152 t. 11, non Kotschy.

I. capitatae similis; staminibus omnibus fertilibus; legumine longioré, glabro, fere semper 2-spermum; indumento ramorum, rhachidum foliorum, costarumque foliolorum sparsiore, patentiore et illo inflorescentiarum calycumque sparsiore differt.

Herba annua, sublignosa, erecta, ramosa, ad 1 m. alta, [radice primaria valide persistente], ramis, ut rhachidibus foliorum, costis foliolorum, inflorescentiis calycibusque, pilis paucis ± patentibus conspersis; ramis superioribus valde angulatis. *Stipulae* angustissime attenuato-deltoideae, persistentes, ad 5 mm. longae. *Folia superiora* 5–7-foliolata, rhachide ad 1·4 cm. longa, angulata, petiolo *c.*4 mm. longo incluso, haud, vel paulo ultra foliola lateralia prolongata. [Folia inferiora interdum ad 13-foliolata, rhachide ad 3 cm. longa, petiolo ad 12 mm. longo incluso]. *Stipelli* filiformes, *c.*0·7 mm. longi; foliola oblanceolata, minute strigulosa, saepius in partibus plantae superioribus, complicata, vel involuta et hinc aspectu angustissima, *c.*1 cm. longa, 2 mm. lata. [Illa foliorum inferiorum ad 1·6 cm. longa, 4 mm. lata.]

Paniculae ramos productos terminantes, dense congestae, subcapituliformes, bracteis inferioribus, paniculas involucrantibus, foliiformibus, pinnatis, superioribus vel pinnatis vel trifidis lobis calycum aspectu simillibus. *Pedicelli c.*1 mm. longi ad, et post, anthesin erecti. *Calyx* sparse patente pilosus, ad basin fissus, lobis anguste lanceolatis, ad anthesin *c.*5 mm. longis, deinde ad *c.*8 mm. accrescentibus. *Vexillum* extus pubescens, anguste cuneato-ellipticum, apiculatum, *c.*4 mm. longum, 1·7 mm. latum. *Alae* glabrae. *Carina* glabra, haud rostrata, anguste et acute calcarata, *c.*3·4 mm. longa. *Stamina* omnia fertilia, antheris apiculatis, basin muticis, *c.*0·5 mm. longis, filamentis 3 mm. longis. *Ovarium* glabrum 2-vel rare 1- ovulatum, 1 mm. longum, stylo glabro 3 mm. longo, 1 mm. infra stigma globosum 0·15 mm. diametro, arcuato. [*Legumen* brunneum glabrum 2- vel rare 1- spermum, ad 4 mm. longum, 1·4 mm. latum, 0·7 mm. crassum, endocarpio haud maculato.]

Nigeria, Old Calabar, *Milne* (1866) s.n. BM; Gaboon, Munda distr. Sibange farm, *Soyaux* 423, Milne s.n. Fr. Congo, Dolo, *Schlechter* 12470 K, BM. Belg. Congo: Bas Congo, Stanley pool *Hens* series B 85 K, BM; Dembo *Sapin* s.n.: Forestier Central, Eala *Louis* 1924, *Corbisier Baland* 1890, *Lejeune* 1, Bombilo *Casteels* 23: Kasai, Mokaba *Vanderyst* 3626 K, BM; Kunzulu *Vermoesen* 2023 BM: Ubangi-Uele, Libenge-Zongo *Lebrun* 1622 BM: Bas Katanga, Kamina *Quarré* 2980: locality unknown *Burton* s.n., *Vermoesen* 2467 K, BM. Uganda U4 Central Bugala Sese, 1200 m., grassland, subshrub 50 cm. tall, dull red flowers, July 1945, *Purseglove* 1705 K holo. EA isotype; ibid up to 1 m. tall, 20.2.1933, *Thomas* 809; ibid, 5.6.1932, *Thomas* 117; Buyama island Sese, frequent in Savannah, Dec. 1922, *Maitland* 358; Bukasa island, Sese, shallow soil on top of hills or cliffs in poor pasture, very common, flowers red, insignificant, 18.7.1951, *Norman* 26. Angola near the Congo river, Muanda river, 50 m., 13.3.1922, *Gossweiler* 8806 K, BM; Sumba Peco, 50 m., Feb. 1926, *Gossweiler* 9184 K, BM; Ambriz *Welwitsch* 2040 BM; without locality *Boss* in TRV 36706 BM.

The fruits, not shown in the type, are described from *Louis* 1924.

Although the two species have not hitherto been separated the 24 specimens listed above differ consistently from the 27 specimens of *I. capitata* which have been examined in the characters listed in the key, with only two exceptions, *Heudelot* 642 from French Guinea and *Mearns* 2763 BM from Kigoma-Butiaba in Uganda. Both of these approach *I. paracapitata* in the calyx indumentum and the latter in addition has a 2-seeded pod. Kotschy described *I. capitata* as having six fertile and four sterile stamens, as is clearly illustrated in his plate. In *I. paracapitata* the stamens, wherever counted, seem to be all fertile. *Mrs. Richards* 5350 from Abercorn agrees with Kotschy's description and plate in having four sterile stamens. On the other hand *Dalziel* 24 from Lokoja, Central Nigeria, which is typical *I. capitata* except in its unusually short calyx lobes, has all 10 stamens fertile.

The two species appear to have different climatic requirements and their ranges, so far as is known, do not overlap. *I. paracapitata* occupies an area of heavy rainfall where the climax vegetation is equatorial rain forest while *I. capitata* occupies, more or less continuously, a belt of somewhat drier country surrounding the area of *I. paracapitata*.

C.2.1.9. **I. congesta** *Welw. ex. Bak.*, F.T.A. **2**, 70 (1871); L.T.A. No. 15; F.C.B. **5**, 151. including var. *flamans* Bak.f., L.T.A. 99 (1926).

Port. Guinea; Sierra Leone; Ivory Coast; Togoland; Nigeria; B. Congo; A.-E. Sudan, *Wyld* 701 BM; Uganda, U2–3, U4 *Lankester* s.n. type of var. *flamans*; Kenya, K4-5, K7; TanganyikaT1, T3-5; Port. E. Afr.; N. Rhodesia; Angola, *Welwitsch* 2041 type.

C.2.a.10. **I. lotononoides,** *Bak.f.*, J. Bot. **41,** 187 (1903); L.T.A. No. 13.

Oubangui Chari; A.E. Sudan, *Schweinfurth* Ser **2**, 63 holotype.

C.2.a.11. **I. dasycephala** *Bak.f.*, L.T.A. 100 (1926) No. 16.

Nigeria, *H.V. Lely* 637 ; N. Cameroons, *P. A. Talbot* Oct. 1910 s.n. BM holotype.

C.2.a.12. **I. brassii** *Bak.*, F.T.A. **2,** 76 (1871); L.T.A. No. 55.

Gold Coast, *Brass* s.n. BM holotype.

The statement in F.T.A. that the specimen comes from Sierra Leone appears to be an error. This is possibly merely an extreme variant of *I. pulchra*.

C.2.a.13. **I. pulchra** *Willd.*, Sp. Pl. **3,** 1239 (1802); F.T.A. **2,** 76 (wrongly attributed to Vahl); L.T.A. No. 60; F.C.B. **5,** 154.

I. rufescens Poir. in Lam. Encyc. Suppl. **3,** 148 (1813).

I. dorycnium Fenzl., Flora **27,** 312 (1844) (nomen).

I. pulchra Willd. var. *andongensis* Hiern., Cat. Welw. Afr. Pl. **1,** 208 (1896).

I. dupuisii Mich., Bull. Soc. Roy. Bot. Belg. **36**: 2, 55 (1897); Ill. Fl. Cong. t. 49; L.T.A. No. 56 p.p. quoad typum solum; non sensu Hutch. et Dalz., F.W.T.A. **1,** 393.

" Guinea" (*unknown coll.* s.n. B holotype).

Senegal; Gambia; Port Guinea; Sierra Leone; Liberia; Ivory Coast; Fr. Sudan; Gold Coast (*unknown Danish coll.* s.n. P. holotype of *I. rufescens*); Togo; Dahomey; Fr. Niger Col., *Hagerup* 487 BM.; Nigeria; Oubangui-Chari; B. Congo, *Dupuis* s.n. BR holotype of *I. dupuisii*; A.E. Sudan, *Kotschy* 576 isotype of *I. dorycnium*; Ethiopia; Tanganyika, T3 *Peter* K399; Angola, *Welwitsch* 2047 BM holotype of var. *andongensis*.

This species varies considerably: occasionally plants are found, e.g. *Meikle* 1185 from Nigeria in which all the leaves seem to be simple: no discontinuity can be seen between these and those, e.g. *Welwitsch* 2047 (var. *andongensis*) in which the leaves have up to 9 leaflets.

C.2.a.14. **I. vanderystii** *Gillett* sp. nov.

Species *I. pulchrae* similis: ab ea indumento sparsiore, calycis lobis latioribus, post anthesin accrescentibus, legumineque glabro, differt.

Herba erecta ramosa ultra 50 cm. alta, ramis ascendentibus, $\pm$ costatis, subsparse $\pm$ appresse pilosis, radice haud visa. *Folia* inferiora *c*.5-jugata, ubique pilis medifixis $\pm$ appressis, ad paginam superiorem *c*.0·4 mm. longis, aliter *c*.0·9 mm. longis, subsparse obsita; stipulae anguste deltoideae vel lineares, brunneae, scariosae, ad 5 mm. longae, basin ad petiolam 1 mm. connatae; rhachis *c*.30 mm. longa, petiolo 2–4 mm. longo incluso, ultra foliola lateralia subopposita *c*.3 mm. prolongata stipellae filiformes, ad 0·7 mm. longae; petioluli *c*.0·7 mm. longi; foliola oblanceolata, apice rotundata, mucronata, ad 16 mm. longa, 5 mm. lata. terminale aliis paulo majore. *Folia* superiora, sive paniculae bracteae, reducta, 3–2–1- jugata vel simplicia vel suprema stipulae solae. *Flores* subumbellatim ad apices ramorum et ramulorum 9–20 mm. longorum subpedunculiformum paniculae dispositi, pedicellis ad 5 mm. longis, *c*.0·5 mm. infra calycem articulatis, ad articulum bracteolis binis linearibus brunneis praeditis. *Calyx* ad anthesin aureo-brunneus, pilis $\pm$ flavis, longis, patulis, subsparse obsitus, fere ad basin fissus, tubo 0·5 mm. longo, lobis lanceolatis pellucide aurantiaco punctatis, margine *c*.0·8 mm. longe laceratis, *c*.3·6 mm. longis 1·1 mm. latis (laceris exclusis); post anthesin accrescens, lobis *c*.4 mm. longis 1·8 mm. latis (laceris exclusis). *Vexillum* extus subfusco strigosum, ellipticum, apiculatum, *c*.3 mm. latum, 5 mm. longum. *Alae* glabrae, late dolabriformes, basin auriculatae, 1·4 mm. latae, 4·5 mm. longae, unguiculo 0·3 mm. longo incluso. *Carina* extus subfusco strigosa, calcare acuto 0·6 mm. longo praedita, haud rostrata, 5·5 mm. longa. *Stamina* omnia fertilia, filamentis 4–5 mm. longis, antheris ellipticis, basin muticis, 0·5 mm. longis, mucrone apicale fusco 0·1 mm. longo incluso. *Ovarium* glabrum brunneum 2-ovulatum, 1·2 mm. longum stylo ad 4·5 mm. longo, *c*.1·5 mm. infra stigma hemisphaericum 0·15 mm. diametro rectangulare curvato. *Legumen* glabrum, oblongum, atro-brunneum, 2-spermum, 4·5 mm. longum, 1·8 latum, 1 mm. crassum.

S.W. BELGIAN CONGO, Kwango diocese, Mfue', March 1914, *H. Vanderyst* 3734 BM holotype.

C.2.a.15. **I. nigricans** *Vahl ex Pers.*, Syn. **2**, 327 (1807), ampl. Bak.f., J.Bot· **41**, 193 (1903); L.T.A. No. 61; non sensu Bak., F.T.A. **2**, 78, nec Hutch. et Dalz., F.W.T.A. **1**, 391.

I. elegans Schum. et Thonn., Beskr. Guin. Pl. 368 (1929); L.T.A. No. 42.

"*I. dupuisii Mich.*" sensu Bak.f., L.T.A. No. 56 p.p. quoad *Dalziel* 598 et sensu Hutch. et Dalz., F.W.T.A. **1**, 393; non Mich.

"*I. aff. I. bracteolata G. et P.*" sensu Chevalier, Expl. Bot. Afr. Occ. Fr. 172.

FR. SUDAN, *Bellamy* 122 P, Koulikoro *Chevalier* 25015 P, Kerekoro *Chevalier* 3220 BM; GOLD COAST, *Thonning*? s.n. (P. holotype) BM fragment, isotype, *Thonning* s.n. C. holotype of *I. elegans*, *Irvine* 735 *Morton* in GC 9255; FR. TOGO, *Mahoux* 228 P; NIGERIA, *Dalziel* 598.

Chevalier 3220 and 25015 differ from the remaining specimens in having more white hairs on the calyx and corolla.

This apparently uncommon species has been much misunderstood. Schumacher gave it a new name, *elegans*, although stating that his specimen was probably a portion of the same gathering which had, in Paris, been described by Persoon. He states that Persoon's description (which is totally inadequate) does not fit his plant but the difference seems to be merely that he describes the leaflets as narrowly cuneate ovate-oblong, Persoon as linear. Baker in the F.T.A. described *Baikie* 20 (*I. latisepala*) from the Niger as *I. nigricans*. His son in 1903, having obtained a scrap of the type of *I. nigricans* from Paris, perceived that it was a totally different plant from *Baikie* 20 and gave it a proper description. Unfortunately he went on to suggest that *Baikie* 20 might be *I. elegans*, with whose description it disagrees in several important respects. In 1926 in the L.T.A. he repeats his former opinions but also identifies *Dalziel* 598 as *I. dupuisii*, which it is not. Finally Hutchinson, in the extreme pressure of work under which he was forced to prepare his F.W.T.A., repeats Baker's mistaken treatment of *I. nigricans*, ignoring E. G. Baker's correction, repeats E. G. Baker's mistaken treatment of *I. dupuisii* and omits *I. elegans* altogether.

C.2.a.16. **I. oubanguiensis** *Tiss.*, Bull. Mus. Hist. Nat. Par. 2nd Ser. **3**, 164 (1931).

"*I. aff. I, parviflora Heyne*" Chevalier, Expl. Bot. Afr. Occ. Fr. 175.

FR. SUDAN, *Chevalier* 24,653 bis; OUBANGUI CHARI, *Chevalier* 6539 (P, syntype) K isotype (*Le Testu* 3645 P, *Tisserant* 734 P syntypes); A. E. SUDAN, *c*.5 km. from Wau, *Hoyle* 375 BM.

There is some doubt about the determination of *Chevalier* 24,653 bis which is very imperfect.

C.2.a.17. **I. leptoclada** *Harms.*, Eng. B. Jahrb. **40**, 38 (May 1907); Bull. Soc. Bot. Fr. **54** mem. **8**, 14 (Aug. 1907); L.T.A. No. 5.

"*I. n.sp. aff. I. procera*" sensu Chevalier, Expl. Bot. Afr. Occ. Fr. 176.

"*I. bracteolata DC.*" sensu Hutch. et Dalx., F.W.T.A. 1, 393 p.p. non DC.

GAMBIA, *Dalziel* 8027; FR. SUDAN, *Chevalier* 651 isotype, *Bellamy* 702 P, *Roberty* 17060 K, G; GOLD COAST, *Lloyd Williams* 116.

C.2.a.18. **I. bracteolata** *DC.*, Prod. **2**, 223 (1825); L.T.A. No. 59.

I. grisea Bak., F.T.A. **2**, 120 (1871); L.T.A. No. 88 pro min. parte non *I. grisea* Desv. (1814).

SENEGAL, *Perrottet* 198 (P, holotype) BM isotype; FR. SUDAN; GOLD COAST; DAHOMEY (fide Tisserant); FR. NIGER COL., *Hagerup* 489 BM; NIGERIA,Abeokuta *Irving* s.n. syntype of *I. grisea* Bak.; A.E. SUDAN, *Kotschy* 553, *Andrews* 104.

This species varies greatly, particularly in the extent to which pinnate or simple leaves predominate. Specimens showing simple leaves only are readily confused with *I. leptoclada*.

C.2.b. subsect. Trichopodae (Bak.) Gillett.

"Group" *Trichopodae* Bak., F.T.A. **2**, 67 emend; Bak.f. L.T.A. 114 p.p.

Lectotype species *I. trichopoda* Leprieur ex Guill. et Perr.

Centre of distribution West Africa.

Lower leaves pinnate, the upper often simple, bract-like. Flowers single on slender glabrescent pedicels 5 mm. or more long, much longer than the flowers themselves. These pedicels are jointed just below the flower, the lower part probably representing the peduncle of a raceme reduced to a single flower. Calyx about as long as corolla, its lobes several times as long as the tube. Fruit short and broad 1–2-rarely 3–4-seeded. Seeds close together, endocarp not spotted.

Trichopodae in the strict sense, as here defined, are related to the **Paniculatae.** *I. monantha*, and *I. uniflora* Ham. from Mysore have a very similar inflorescence, but the flowers and fruits are those of **Dissitiflorae.**

Pod less than 1·5 mm. wide; more than 4 (*c.*8–10)-seeded, *c.*10 mm. long with spotted endocarp; calyx half as long as corolla, the lobes about as long as the tube . . . . *monantha.*

Pod 1·5 mm. wide; 1–2, rarely 3–4-seeded, less than 10 mm. long, endocarp not spotted; calyx ± as long as corolla, its lobes much longer than the tube:

Calyx lobes narrowing from the base, where they are widest:

Slender annual up to *c.*30 cm. tall; indumentum ± spreading . . . *I. trichopoda.*

Leaflets elliptic oblong, fruits 1–2-seeded . . . . α. var. *trichopoda.*

Leaflets suborbicular, fruits *c.*4-seeded . . . . β. var. *oubanguiensis.*

Stiff erect bushy herb up to 2 m. tall; leaflets narrow oblong or oblanceolate; indumentum closely appressed . . . . 2. *nigritana.*

Calyx lobes narrowing from just below the middle, where they are wider than at the base; indumentum long, spreading . . . . 3. *latisepala.*

C.2.b.1. **I. trichopoda** *Leprieur ex Guill. et Perr.*, Fl. Seneg. 177, t. 47 (1832), L.T.A. No. 68.

α. var. **trichopoda.**

SENEGAL, *Leprieur* s.n. (P, holotype) BM, isotype *Heudelot* 272, *Thierry* 223, *Berhaut* 710 P.

β. var. **oubanguiensis** Tiss., Bull. Mus. Hist. Nat. Par. 2nd ser. **3**, 165 (1931).

OUBANGUI-CHARI, *Le Testu* 3400 P syntype, (*Tisserant* 1293 P 2276 P syntypes).

C.2.b.2. **I. nigritana** *Hook.f.* in Hook. Niger Flora 294 (1849); L.T.A. No. 58; F.C.B. **5**, 151.

I. oligantha Harms ex Bak.f., L.T.A. 111 (1926), No. 57.

32

SENEGAL (fide Tisserant); FR. SUDAN, *Davey* 48, *Chevalier* 3235; GOLD COAST; NIGERIA, *Vogel* s.n. holotype; FR. CONGO, *Schlechter* 12666 holotype of *I. oligantha*; OUBANGUI-CHARI, *Le Testu* 2333 BM; B. CONGO (fide Cronquist); A.E. SUDAN, *Brown* s.n. (a poor specimen: identification doubtful).

Schlechter 12666 has a different aspect from the other material, having shorter internodes and fewer compound leaves: this is probably merely the result of different conditions of growth.

C.2.b.3. **I. latisepala** *Gillett*, Kew Bull. **1955**, 575 (1956).

"*I. nigricans Vahl?*" sensu Bak., F.T.A. **2**, 78; Hutch. et Dalz. F.W.T.A. **1**, 391: non Vahl ex Pers. Syn. **2**, 327 (1807), nec DC. Prod. **2**, 230.

NIGERIA.

Section C.3. **Indigofera.**

Subsection C.3.a. **Juncifoliae** *Harvey*, Fl. Cap. **2**, 163 (1862); L.T.A. 101.
Lectotype species *I. filifolia* Thunb.
Coast of South East Africa.

Petioles many times longer than the small terminal leaflet, persistent when the latter falls off. Fruiting pedicels erect or spreading. Endocarp spotted, seeds $\pm$ separated.

C.3.a.1. **I. podophylla** *Benth. ex Harv.*, Fl. Cap. **2**, 168 (1862); Ic. Pl. t. 1800; L.T.A. No. 17.

PORT E. AFR. coast from 21° to 26° 30′ S., Delagoa Bay *Forbes* s.n. holotype, etc.

C.3.b. **Brevi–erectae** subsect. nov.

Type species, *I. simplicifolia* Lam.
Centre of distribution, A.-E. Sudan-Angola.

Folia simplicia vel pinnata. Inflorescentiae racemi axillares; pedunculi breves, saepius axe florifero breviores, haud tenues nec filiformes; pedicelli fructiferi leguminaque erecti, vel exteriores $\pm$ patentes. Calyx vel brevis vel longus. Legumina diversa, 1–9-sperma, recta. Endocarpium diversum. Semina inter se haud separata.

Leaves simple:
Pods 1–2-seeded:
    Pods 1-seeded, $\pm$ spherical; endocarp not spotted; leaves linear . . . 1. *linifolia*.
    Pods 2-seeded, oblong; endocarp spotted; leaves oval . . . . . . 2. *cordifolia*.
Pods 3- or more seeded, elongate:
    Leaves up to 3 mm. wide; calyx teeth *c.*1 mm. long; pod 1·8–2·5 mm. wide; endocarp minutely spotted; seeds hardly pitted:
        Hairs on upper surface of leaf set at an angle of 80°–90° to the midrib; pod somewhat flattened, *c.*2·5 mm. wide 3–6-seeded; persistent style base *c.*0·3 mm. long . . . . 6. *tetrasperma*.
        Hairs on upper surface of leaf set at an angle of *c.*45° to the midrib; pod $\pm$ tetragonal, *c.*1·8–2·0 mm. wide, usually 8–12-seeded; persistent style base *c.*1·1 mm. long . . . . 7. *simplicifolia*.
    Leaves up to 10 mm. wide; calyx teeth *c.*5 mm. long; pod *c.*4 mm. wide; endocarp strongly spotted; seeds conspicuously pitted . . . . 8. *leprieurii*.

Leaves compound:

Leaflets always, or often, alternate; inflorescence very dense, sessile or shortly pedunculate; fruits at least 1·5 mm. wide, not above 12 mm. long:

Stipules ± triangular, brownish, scarious, glabrescent; pod *c.* 2-seeded somewhat tetragonal;endocarp not spotted; seeds not pitted . . . 3. *enneaphylla.*

Stipules narrow-lanceolate, greenish, thickly strigose:

Pod *c.*2 mm. wide, somewhat tetragonal, not torulose, 2 seeded; endocarp spotted; seeds with 4 pits on each face . . . .   4. *trigonelloides.*

Pod *c.*1·5–2 mm. wide, not tetragonal, torulose, 3–6 seeded; endocarp not spotted; seeds not pitted . . . . 5. *sessiliflora.*

Leaflets not alternate:

Calyx lobes as long as tube, or longer, or, if shorter, then the leaflets much less than 3 cm. long and 1 cm. wide:

Pod more than 12 mm. long; leaf rhachis prolonged well beyond lateral leaflets:

Pod more than 2 mm. wide; leaflets usually over 12 mm. long, often over 2 mm. wide:

Leaflets usually more than 11, often 15–23, about 2–5 times as long as wide; petiole usually under 8 mm. long; pods ± winged; hairs on calyx long (*c.*1 mm.) ± spreading; flowers erect in bud, later reflexed and finally fruits becoming erect again; endocarp conspicuously spotted . . . . 9. *prieureana.*

Leaflets usually fewer than 11, 4–10 times as long as wide; petiole usually over 10 mm. long; pod not winged; endocarp spotted . . . 10. *stenophylla.*

Hairs on calyx *c.*0·5 mm. long, ± appressed; flowers erect throughout . . . . α. var. *stenophylla.*

Hairs on calyx *c.*1 mm. long, ± spreading; flowers reflexed after opening and later again becoming erect . . . . . β. var. *ampla.*

Pod 1·7 mm. wide or less:

Glandular hairs present; leaflets up to 11, rarely more than 13 mm. long and 4 mm. wide; petiole 3–5 mm. long . . . .
colutea var. *dembianensis.*

Glandular hairs absent; leaflets often more than 11, rarely more than 12 mm. long or 2 mm. wide; petiole rarely as much as 3 mm. long
subargentea.

Pod 12 mm. long or less, or, if more then leaf rhachis often not prolonged beyond lateral leaflets:

Pod *c.*10 times as long as wide (*c.*9–12 mm. long, 1 mm. wide), endocarp not spotted; infructescence shorter than subtending leaf; leaflets 3–7; petiole 0–2 mm. long; plant usually ± green . . . 11. *griseoides.*

Pod not more than 4 times as long as wide, or, if so, then infructescence longer than subtending leaf:

Leaflets, at least of the lower leaves, often more than 7, rhachis nearly always prolonged beyond lateral leaflets:

Hairs all closely appressed; inflorescence always shorter than subtending leaf; calyx lobes about as long as tube; suffruticose; endocarp spotted . . . . 12. *mildbraediana.*

Hairs on young stems and inflorescences ± spreading; inflorescence often longer than subtending leaf; calyx lobes much longer than tube:

Annual; calyx $\frac{3}{4}$ or $\frac{4}{5}$ of length of stamens; upper leaves not
shorter, nor with fewer leaflets than the lower; petiole not less
than 4 mm. long . . . . .13. *andrewsiana.*

Suffruticose perennial; calyx lobes $c.\frac{1}{2}$ as long as stamens; upper
leaves much shorter and fewer-foliolate than the lower;
petiole often as short as 2 mm. . . . . *subulifera.*

Leaflets not more than 7, rhachis often not prolonged beyond lateral
leaflets:

Calyx lobes about as long as tube; plant silvery; hairs soft, often
spreading; fruit 1-, rarely 2–3-seeded . . . . . . 14. *microcalyx.*

Calyx lobes *c.*3 times longer than tube; medifixed hairs stiff, appressed;
fruit several-seeded:

Leaves subtending inflorescences usually 1–2-jugate . . . . *hilaris.*

Leaves subtending inflorescences 3-jugate . . . . *heterocarpa.*

Calyx lobes much shorter than the tube; leaflets 7, *c.*3 cm. long and 1 cm. wide;
standard densely covered outside by a golden brown appressed indumentum . . . . 15. *pobeguinii.*

C.3.b.1. **I. linifolia** (*L.f.*) *Retz.*, Obs **4,** 29 (1786); **6,** 33, t. 2. (1791); L.T.A.
No. 4.

Hedysarum linifolium L.f., Suppl. Plant. 331 (1781).

Sphaeridiophorum linifolium (L.f.) Desv., J. Bot. **1,** 125, t. 6 (1813).

S. abyssinicum Jaub. & Spach., ı11. Pl. Or., t. 494 (1857).

Afghanistan; W. Pakistan; India, *unknown collector* BM, type; E. Pakistan;
Siam; China (Yunnan); Indonesia; Australia; A.E. Sudan; Eritrea; N.
Ethiopia (*Schimper* (Aug. 1852) s.n. P. holotype of *S. abyssinicum*).

The apparent absence of this species and the next from Tropical Arabia is
difficult to understand.

C.3.b.2. **I. cordifolia** *Heyne ex Roth.*, Nov. Pl. Sp. 357 (1821); L.T.A. No. 18.

Afghanistan; Baluchistan; W. Pakistan; India, *Heyne* s.n. type; Timor;
Cape Verde Islands; Mauritania, *Monod* in IFAN 10986 IFAN; Fr. Niger
Col., Aïr 17°15′N *Chopard & Villiers* s.n. IFAN; A.E. Sudan; Eritrea; N.
Ethiopia; Socotra.

C.3.b.3. **I. enneaphylla** L., Mant. **2,** 272 (1771), non sensu Bak., F.T.A. **2,** 95,
nec. sensu Ecklon & Zeyher, Enum. Pl. Afr. Austr. 244 No. 1621.

W. Pakistan; India; Burma; Ceylon; Siam; Indochina; Indonesia; New
Guinea; Australia.

Welwitsch 2011, 2011b cited as *I. enneaphylla* by Baker are *I. microcarpa* Desv.,
Ecklon & Zeyher 1621 is *I. alternans* DC. fide Harvey in Flor. Cap.

C.3.b.4. **I. trigonelloides** *Jaub. & Spach.*, Ill., Pl. Or. 5, t. 482 (1856); Bak.
F.T.A. **2,** 79 pro maj. parte; L.T.A. No. 75: non sensu Bak., Fl. Br. Ind. **2,** 94.

I. aeruginis Schwnf., Beitr. Fl. Aeth. 11 (1867).

Ethiopia, nr. Gageros 12.9.1854, *Schimper* 2272 isotype of *I. trigonelloides* and
I. aeruginis. S.W. Afr., Tsumeb, 19 & 23.3.1934, *Dinter* 7344, 7344A.

The plants from S.W. Africa are larger than the type, and 7344A in particular
has much narrower leaflets, but they seem conspecific.

C.3.b.5. **I. sessiliflora** *DC*, Prod. **2,** 228 (1825); L.T.A. No. 76.

I. tribuloides Boiss., Fl. Or. **2,** 189 (1872).

"I. arabica Jaub. & Spach" sensu Bak., F.T.A. **2,** 103 non Jaub. & Spach.

"I. trigonelloides Jaub. & Spach" sensu Bak., Fl. Br. Ind. **2,** 94 non Jaub. & Spach.

SAUDI ARABIA, 19°N, 41°10′E, *Vesey Fitzgerald* 16476/5 BM; N.W. PUNJAB., *Griffiths* 1174 isosyntype of *I. tribuloides*, etc.; SIND, *Stocks* s.n. ? isosyntype of *I. tribuloides*; MAURITANIA; SENEGAL, *Perrottet* 183 (P, holo.) BM isotype, etc.; FR. SUDAN, *Chevalier* 1266; FR. NIGER COL., *De Wailly*, 5242 P, 5258 P. A.E. SUDAN, *Petherick* s.n. *Harrison* 940; *Newberry* 52 BM; ERITREA, *Bally* 6893, *Popov* 1421.

In its fewer and larger leaflets *Harrison* 940 approaches *I. trigonelloides*. Boissier gives "Afghanistan" as the area for his *Griffiths* and *Stocks* types of *I. tribuloides*. This seems to be in error for the Indus valley.

C.3.b.6. **I. tetrasperma** *Vahl ex Pers.*, Syn. **2,** 325 (1807); ampl. DC., Prod. **2,** 222 et Schum. & Thonn., Beskr. Guin. Pl. 365 (1829); sensu Bak., F.T.A. **2,** 72 et Bak.f., J. Bot. **41,** 189 partly, non quoad plantam Heudelotii; L.T.A. No. 24.

"I. simplicifolia Lam." sensu Hutch. & Dalz., F.W.T.A. **1,** 393 partly, non Lam.

"I. procera Schum. & Thonn." sensu Chevalier, Expl. Bot. Afr. Occ. Franc. 176 partly, non Schum. & Thonn.

IVORY COAST, Cercle de Mankono, Buanolougou-Marabasiassa, *Chevalier* 22007, Nzi Valley Baoule, *Chevalier* 2210; GOLD COAST, N.W. Ashanti, Sunyani *Vigne* 2475, coastal area near Accra and Achimota, *Irvine* 128, 570, 792, *Ankarah* in GC 20025, (*Thonning* s.n. P holo. C isotype) Photo of the C sheet at BM; TOGO *Büttner* 122 BM.

The name *I. tetrasperma* has been misapplied from the beginning. Guillemin, Perrottet and Richard, Fl. Seneg. 183 (1832) state that Vahl himself labelled specimens of *I. senegalensis* Lam. in the herbaria of Desfontaines and De Jussieu as *tetrasperma*. Ignoring Persoon's description they therefore state that *I. tetrasperma* Vahl is a synonym for *I. senegalensis*; while recognising that *I. tetrasperma* sensu Schum. & Thonning is a different species. Baker, F.T.A. **2,** 72 states that Webb "examined an authenticated specimen" (of *I. tetrasperma* Pers.) "in the herbarium of Desfontaines, this latter is *I. senegalensis*". On this ground he concludes that *I. tetrasperma* Schum. & Thonn. is not the same species as *I. tetrasperma* Vahl ex Pers. However, the fact that Vahl seems to have applied his epithet to more than one species is irrelevant to the question at issue, which is "what is Persoon's type?" Persoon cannot possibly have been describing a specimen of *I. senegalensis* for he describes the leaves as simple and the plant as virgate, neither of which statements apply to *I. senegalensis*, while both apply to the present species. Furthermore De Candolle, Prod. **2,** 222 states that he has examined Persoon's type, that this was collected by Thonning in Guinea (i.e. the Gold Coast or Togoland, a region where *I. senegalensis* has never been collected), that it is distinct from *I. senegalensis* and that it has "floribus axillaribus subsolitariis" which fits our present plant (and very few others) but certainly not *I. senegalensis*. Thus there is no real reason to doubt that Schumacher was right in thinking that he was amplifying Persoon's very inadequate description, and, in all probability, the plant which he described and that described by Persoon were both portions of one gathering made by Thonning. Baker did not see Thonning's plant and identified *Heudelot* 196 as *I. tetrasperma* because the few pods which remain on it are *c.*4-seeded. However, as recognised by Tisserant and others, *Heudelot* 196 and several other sheets distributed as *tetrasperma* (e.g. *Schlechter* 12477 from the Congo) are nothing but *I. simplicifolia* in which species it is fairly common for a few of the pods to be much shorter than usual. In the Gold Coast, however, the locus classicus for *I. tetrasperma*, all the material (cited above) formerly identified as *I. simplicifolia* proves to differ from that species in the characters listed below, and there can be no doubt that this is the true *tetrasperma*.

I. tetrasperma	*I. simplicifolia*
1. Hairs on upper surface of leaf dense, set at an angle of 80–90° with the midrib.	Hairs on upper surface of leaf less dense, set at an angle of *c.*45° with the midrib.
2. Pod $\pm$ flattened, 10–15 mm. long, *c.*2·5 mm. wide, *c*1·0 mm. thick.	Pod cylindrical-subtetragonal usually 15–20 mm. long, sometimes shorter, *c.*1·8–2·0 mm. wide and 1·4 mm. thick.
3. Apex of pod relatively blunt, the 2 sides converging at *c.*90°.	Apex of pod less blunt, the 2 sides converging at *c.*60–70°.
4. Persistent style base *c.*0·3 mm. long.	Persistent style base *c.*1·1 mm. long.
5. Seeds 3–6.	Seeds 8–12, occasionally fewer.
6. Seeds *c.*1·5 mm. long, 1·5 mm. wide, 0·8 mm. thick, elliptic in cross section.	Seeds *c.*1 mm. long, 1 mm. wide, 0·8 mm. thick, rhomboidal in cross section.

C.3.b.7. **I. simplicifolia** *Lam.*, Encycl. **3**, 251 (1789); L.T.A. No. 25; F.C.B. **5**, 129.

"*I. tetrasperma Schum. & Thonn.*" sensu Bak., F.T.A. **2**, 72 (1871) non Schum. & Thonn. nec Vahl ex Pers.

I. tetrasperma Vahl ex Pers. var. *hexasperma* Vatke, Oest. Bot. Zeit. **28**, 200 (1878).

I. polysperma De Wild. & Th. Dur., Bull. Herb. Boiss. 2nd Ser. **1**, 14 (1901).

SENEGAL; FR. GUINEA; SIERRA LEONE, *Smeathman* s.n. BM isotype; LIBERIA; NIGERIA; GABOON (fide Tisserant); FR. CONGO; B. CONGO, *Gillet* s.n. BR holotype of *I. polysperma*; OUBANGUI CHARI; A.E. SUDAN; UGANDA, U1, U3–4; ZANZIBAR, *Hildebrandt* 939 BM isotype of var. *hexasperma*; PORT. E. AFR.; ANGOLA.

C.3.b.8. **I. leprieurii** *Bak.f.*, J. Bot. **41**, 190 (1903); L.T.A. No. 26; F.C.B. **5**, 130.

I. macrocarpa Leprieur ex Bak., F.T.A. **2**, 72 (1871); Chevalier, Expl. Bot. Afr. Occ. Fr. 175: non Desvaux (1826).

"*I. simplicifolia Lam.*" sensu Guill. & Perr., Fl. Seneg. 173 et Hutch. & Dalz., F. W. T. A. **1**, 393 pp.; non Lam.

SENEGAL, *Leprieur* s.n. (P syntype) BM isosyntype (*Perrottet* s.n. P syntype); GAMBIA; PORT GUINEA; SIERRA LEONE: FR. SUDAN (fide Chevalier); GOLD COAST; DAHOMEY (fide Tisserant); NIGERIA, *Vogel* 58, 94 syntypes *Barter* s.n., 1079, 1229 syntypes; B. CONGO (Bas Katanga fide Cronquist).

C.3.b.9. **I. prieureana** *Guill. & Perr.*, Fl. Seneg. 187 (1832); L.T.A. No. 126.

I. komiensis Tiss., Bull. Mus. Hist. Nat. Par. 2nd Ser. **3**, 170 (1931).

I. stenophylla Guill. & Perr. var. *latifolia* A. Rich., Tent. Fl. Ab. **1**, 179 (1847).

I. stenophylla var. *brachypoda* Steud. ex Bak., F.T.A. **2**, 74 (1871).

"*I. aff.sutherlandioides*" sensu Chevalier, Expl. Bot. Afr. Occ. Fr. 178.

SENEGAL (*Leprieur* s.n. P holotype) *Heudelot* s.n.; FR. SUDAN, *Chevalier* 3226 isosyntype of *I. komiensis*; FR. NIGER, *Money-Kyrle* Fe 13; N. NIGERIA, *Moiser* 4, 95, 267, *Sampson* 19, *Shaw* 100, *Dalziel* 27, 325, 622, *Lely* 507; BRIT. CAMEROONS, North, *McClintock* 13; CHARI (*Chevalier* 9387P, 9425P, syntypes of *I. komiensis*);

A.-E. Sudan, El Obeid Kordofan *Pfund* 20, Gallabat *Schweinfurth* 1816 1816 BM; Ethiopia, North, *Schimper* (1841) 1468, 1711, isotypes of *I. stenophylla* vars. *latifolia* and *brachypoda*.

C.3.b.10. **I. stenophylla** *Guill. & Perr.*, Fl. Seneg. 188, t. 48 (1832); Chevalier, Expl. Afr. Occ. Fr. 177; L.T.A. No. 125; F.C.B. **5**, 160.

α. var. **stenophylla**.

Including var. *macrocarpa* Guill. & Perr. ibid.

Senegal, Bakel (*Leprieur* s.n. P holotype); Gambia (*Perrottet* s.n. P type of var. *macrocarpa*); Fr. Guinea; Fr. Sudan (fide Chevalier); Gold Coast; Nigeria; Oubangui Chari; B. Congo (fide Cronquist); A.-E. Sudan; Uganda, U1–4; Tanganyika, T1.

β. var. **ampla** *Sprague*, Kew Bull. **1909**, 185.

Fr. Sudan, *Scäetta* 3212; Gold Coast, *Morton* in GC 25291; Nigeria, *Dodd* 427 holotype, etc.

I. stenophylla is evidently closely related to *I. prieureana* with which it probably hybridises. Var. *ampla* may well be of such hybrid origin: it has the curious character of *I. prieureana* that the pedicels which are deflexed immediately after flowering subsequently straighten out again, so that the fruits end up erect.

C.3.b.11. **I. griseoides** *Harms* in Warburg & Baum Kunene-Zambesi Exp. 257 (1903); L.T.A. No. 80.

Tanganyika, T5 *Burtt* 5069, T7 *Burtt* 6227; N. Rhodesia; Angola, *Baum* 745 (B† holotype) photo K.

Senni 243 Fi recorded by Chiovenda, Flor. Som. **2**, 159 as this species is *I. colutea*.

C.3.b.12. **I. mildbraediana** *Gillett*, Kew Bull. **1955**, 575 (1956).

"*I. subulifera Welw.*" sensu Tisserant, Bull. Mus. Hist. Nat. Par. 2nd ser. **3**, 167 (1931) non Welw. ex Bak.

Nigeria; Fr. Congo; Gaboon; Oubangui-Chari; A.-E. Sudan; Tanganyika; Angola, *Welwitsch* 2048b p.p. BM iso, P holotype.

C.3.b.13. **I. andrewsiana** *Gillett* sp. nov.

Species *I. stenophyllae* affinis; leguminibus brevioribus, foliolis brevioribus, indumento argentiore, patentiore, differt.

I. mildbraedianae quoque affinis: habitu annuo, leguminibus longioribus, indumento patentiore differt.

Herba annua erecta *c*.45 cm. alta, ramis rigidis angulatis, argenteo pilosis, pilis, praesertim in partibus junioribus, ± patentibus. *Folia* 11–13-foliolata, ubique argenteo-strigosa, *c*.40–50 mm. longa, petiolo 4–6 mm. longo et foliolo terminale *c*.14 mm. longo inclusis; rhachis saepius 1–2 mm. ultra foliola lateralia prolongata; stipulae subulatae, *c*.4 mm. longae, argenteo strigoso-pilosae, pilibus ± patentibus; foliola anguste oblongata vel oblanceolata, apice subacuta, *c*.10 mm. longa, 2 mm. lata. *Racemi* in axillis rigide erecti *c*.10–20-flori, ad 5 [6] cm. longi, sessiles vel ad 6 mm. pedunculati, ubique dense strigoso-pilosi, pilis subpatentibus, floribus leguminibusque erectis; bracteae erectae anguste lanceolatae, *c*.1·5 mm. longae, mox deciduae; pedicelli *c*.0·75 mm. longi. *Calyx* argente strigoso-pilosus, pilis ± patentibus, *c*.4 mm. longus, tubo 1 mm. longo incluso, lobis subaequalibus vel lobo inferiore quam alii ad 0·5 mm. longiore, angustissime triangularibus. [*Corolla* mox decidua. *Vexillum* anguste obovatum, basin attenuatum, extus dense appresse argenteo-strigosum, *c*.4·5 mm. longum 2 mm. latum. *Alae* glabrae 0·2 mm. unguiculatae, auriculatae, dolabriformes, *c*.3·5 mm. longae. *Carina* extus strigosa, calcare

1 mm. longo, subrostrata, *c*.4·5 mm. longa.] *Stamina* omnia fertilia, filamentis *c*.4 mm. longis, [antheris late ellipticis, basin muticis, *c*.0·5 mm. longis, mucrone apicale flavo 0·1 mm. longo incluso.] *Ovarium c*.1·3 mm. longum, appresse pilosum, *c*.4-ovulatum, stylo *c*.4 mm. longo, infra mediam piloso. superne glabro, *c*.1·3 mm. infra stigma capitatum rectangulare curvato. *Legumen* subcylindricum argenteo-strigoso-pilosum, pilis paulo patentibus *c*.0·5 mm. longis, 2–4-spermum, 6–7 mm. longum, 2 mm. latum. *Endocarpium* ad septam inter semina et aliter in linneis quatuor nigro-punctatum. *Semina* brunnea, cuboidea, quoque facie 3–4-foveolata.

A.-E. SUDAN, Lokwara, 33°27′E, 4°40′N, *c*.700 m., on black cotton soil with stones, not eaten readily by stock, Taposan name NYAMARET, 29.9.1953, *A. W. Peers* 2 holotype. Malakal, 31°40′E, 9°30′N, 380 m., a weed, flowers bright pink, 22.1.1930, *N. Douglas Simpson* 7441 K, BM. Sobat, Nile (near Malakal) in cultivated land, mainly eaten by sheep and goats, though cattle will also eat it. *E. Evans Pritchard* 9. UGANDA, U.1. West Nile distr. Pokwach, 31°35′E, 2°30′N, 750 m. sticky black soils of grassy plains, herb 35 cm. tall with minute pink flowers, July 1940, *J. W. Purseglove* 975 K, EA.

As the type does not show the mature corolla this and the anthers have been described from *Simpson* 7441. *Purseglove* 975 varies in having a calyx only 2·5 mm. long.

This species is named after Mr. F. W. Andrews, author of "The flowering plants of the Anglo-Egyptian Sudan".

C.3.b.14. **I. microcalyx** *Bak.*, Kew Bull. **1897,** 256; L.T.A. No. 7; F.C.B. **5,** 155.

I. karongensis Bak., Kew Bull. **1897,** 255; L.T.A. No. 6.

I. bagshawei Bak.f., J. Linn. Soc. **37,** 142 (1905); L.T.A. No. 208.

I. condensata De Wild., Bull. J. Bot. Et Brux. **8,** 148 (1923); L.T.A. No. 84.

FR. CONGO, Brazzaville, *Chevalier* 27767 P; B. CONGO, Leopoldville, *Bequaert* 7201 BR holotype of *I. condensata*, etc.; UGANDA, U3 *Wood* 756, U4 *Maitland* 328, *Bagshawe* 11 BM holotype of *I. bagshawei, Bagshawe* 1413 BM; TANGANYIKA, T1 *Ritchie* H67/35 EA, *Haarer* 2068, T4 *Burtt* 6143, T6 *Schlieben* 2421; PORT. E. AFR., 13–14°S *c*.1200 m. *Johnson* 75; NYASALAND, *Whyte* s.n. holotype, Whyte s.n. holotype of *I. karongensis*, etc.; N. RHODESIA, *Bullock* 3821, *Burtt* 6097, etc.; ANGOLA, nr. Congo, Gossweiler 8597 BM.

This species seems usually to grow on the sandy banks of lakes and, less often, rivers. The material from Lake Nyasa has fruits, 1-, rarely 2- seeded, that from other areas has fruits 1–4-seeded but no clear line can be drawn between the two forms.

C.3.b.15. **I. pobeguinii** *Gillett*, Kew Bull. **1955,** 577 (1956).

FR. GUINEA, *Pobéguin* 976 P. holotype.

Although placed here in the **Brevi-erectae** on account of its ovaries being erect after flowering, this species seems to be closest to *I. emarginella*.

Subsection C.3.c. **Anomalae** subsect. nov.

Planta omnino glabra, corolla incluso, pilis perpaucis medifixis sparsis ad rhaches foliorum solis exceptis. Folia pinnata. Racemus gracilis dissitiflorus, pedicellis fructiferis erectis. Calyx profunde fissus. Legumen ± compressum, 1-spermum.

C.3.c.1. **I. scarciesii** *Scott-Elliot*, J. Linn. Soc. **30,** 76 (1894); L.T.A. No. 9.

FR. GUINEA near Buya Buya, Scarcies river, 12°38′W, 9°45′N on laterite road, 6.2.1892, *Scott-Elliot* 4765 K, BM isotypes. Diakrané, Kankan, decoctions used against fever, *Brossart* in *Chevalier* 11673 P., Santa-Timbo *Chevalier* 12590 P.

A very distinct plant of uncertain affinities. The delicate shoots spring from a persistent woody rootstock. The leaflets have curious translucent lateral veins of a type not seen elsewhere in *Indigofera*.

Subsect. C.3.d. Dissitiflorae (*Bak.*) *Gillett*.

"Group" *Dissitiflorae*., Bak., F.T.A. **2**, 67 p.p. emend.; L.T.A. No. 117.
Lectotype species *I. geminata* Bak.
Centre of distribution E. Tropical Africa.

Leaves, or at least the lower ones, pinnate, the leaflets rather small (rarely above 15 mm. long) opposite. Raceme open, usually few-flowered; the peduncle stiff, wiry, slender, bracts deciduous, fruiting pedicels and fruits stiffly erect or sometimes the lower ones only ± spreading. Calyx shortly or deeply divided, half as long as stamens or less. Standard pubescent. Fruit more than 3-seeded, always straight, endocarp spotted (sometimes rather obscurely so) seeds usually closely pressed together.

Besides the following species, others occur in India, South Africa and Madagascar.

Leaf rhachis prolonged beyond distal leaflet pair:
Fruiting pedicel over 2 mm. long:
Fruits appressed-strigulose, usually over 17 mm. long:
Leaflets 11–31, ± oblong, *c*.2–4 times as long as wide; endocarp distinctly spotted, seeds pitted . . . . 1. *dendroides*.
Leaflets 5–9; endocarp obscurely spotted:
Leaflets cuneate-oblong, *c*.2–4 times as long as wide . . 2. *sp. aff. 1. amitina*.
Leaflets linear *c*.6–15 times as long as wide . . . . 3. *filipes*.
Fruits glabrous or almost so, less than 17 mm. long; endocarp obscurely spotted:
Raceme 6–12-flowered, its axis pubescent; stems ± pubescent, leaflets 7–13
4. *heudelotii*.
Leaflets rarely over 8 mm. wide; peduncles rarely over 15 mm. long; leaves subtending the inflorescences usually shorter than lower leaves (1–3 instead of 4–6-jugate) . . . . α. var. *heudelotii*.
Leaflets up to 12 mm. wide; peduncles up to 30 mm. long; leaves subtending inflorescences not less than 3-jugate, much like lower leaves
β. var. *fairchildii*.
Raceme 3–5-flowered, its axis (with peduncle) under 15 mm. long, stems subglabrous, the few hairs appressed, leaflets 3–9 . . . . . . 5. *elliotii*.
Fruiting pedicel less than 2 mm. long:
Leaflets having a blister at each side visible from above and a third beneath the tip; hairs appressed . . . . 6. *brevicalyx*.
Leaflets without blisters:
Peduncle always bearing but 1 flower; leaves 1–5-foliolate; stipules *c*.4–5 mm. long . . . . 7. *monantha*.
Peduncles bearing 2 or more flowers, or, rarely, a few uniflorous:
Hairs on young stems (or many of them) spreading:
Upper leaves often simple, resembling bracts, peduncle 1–2-flowered
8. *cuneata*.
Upper leaves not simple:
Peduncle 5–9 mm. long; fruit less than 10 mm. long; stipules 4–6 mm. long; margins of young leaflets glandular:

40

Lower leaves with up to 11 leaflets, leaflets lanceolate, acute . . .
9. *manyoniensis.*

Lower leaves with *c.*5 leaflets oblong or obovate, obtuse . . . . . .
10. *micropetala.*

Peduncle 9–27 mm. long:
Stipules *c.*2 mm. long; margins of young leaflets glandular; fruit usually subglabrous, 10 mm. long or more:
Inflorescence axis, with peduncle, 10–15 mm. long, 1–4 (usually 2–3)-flowered; leaflets more than twice as long as wide:
Perennial; pod ± cylindric; contiguous faces of seeds rhombic, *c.*1·1 × 1·0 mm., not grooved . . . . . 11. *tanganyikensis.*
Leaflets 7–15, not pendulous . . . . . α. var. *tanganyikensis.*
Leaflets 3–5, often pendulous . . . . β. var. *paucijuga.*
Annual; pod flattened; contiguous faces of seeds elliptic, *c.*1·0 × 0·7 mm., grooved . . . . 12. *mwanzae.*
Inflorescence axis, with peduncle, 20–45 mm. long, 3–15-flowered; leaflets less than twice as long as wide; perennial
13. *elwakensis.*

Stipules 3–5 mm. long; leaflet margins not glandular; fruit pubescent:
Petiole 1–2 mm.; fruit 14–20 mm.; leaves subtending inflorescences 5–9-foliolate . . . . 14. *ambelacensis.*
Petiole 3–6 mm.; fruit 7–8 mm.; leaves subtending inflorescences 3–5-foliolate . . . . 15. *kongwaensis.*

Hairs on young stems appressed:
Petiole usually under 1 mm. long, at least in the upper leaves:
Leaflets with glandular margins, usually 4 or more times as long as broad; fruits glabrous 13–20 mm. long, *c.*1·3 mm. wide:
Inflorescence 1–3-flowered; annual . . . . 16. *wituensis.*
Leaflets glabrescent above, up to 12 mm. or more long . . . . .
α. var. *wituensis.*
Leaflets finely appressed strigose above, up to 7 mm. long . . . .
β. var. *occidentalis.*
Inflorescence 4–8-flowered; perennial . . . . 17. *mendonçae.*
Leaflets without glandular margins, appressed pilose above, *c.*3 times as long as broad:
Inflorescence 1–2-flowered; fruits *c.*9–11 mm. long, 1·7 mm. wide; annual . . . . 18. *sparsa.*
Inflorescence 4–6-flowered; fruits 17–25 mm. long; perennial
21. *vohemarensis.*

Petiole always at least 1 mm. long, usually more; margins of young leaflets not glandular:
Leaflets 4–5 times as long as wide:
Petiole 1–3 mm. long, usually shorter than stipules . . . 19. *tenuis.*
Pods not over 15 mm.; leaflets not over 10 mm. long . . . . . .
α. ssp. *tenuis.*
Pods often over 20 mm.; leaflets often over 15 mm. long . . . . .
β. ssp. *major.*
Petiole 5–10 mm. long, longer than stipules . . . . 20. *dissitiflora.*

Leaflets less than 3 times as long as wide:

Fruit 18–25 mm. long, 1·7–2 mm. wide; leaflets ± strigose above; strong smell of coumarin; inflorescence 20–25 mm. long, 4–6-flowered; endocarp with intense rather small spots . . . .
21. *vohemarensis*.

Fruit 13–18 mm. long, *c*.1 mm. wide; leaflets glabrous or glabrescent above; plant not smelling of coumarin; inflorescence not above 18 mm. long, 2–4-flowered; endocarp obscurely spotted
22. *congolensis*.

Stipellae well developed, leaflets usually more than 3 . . . . . .
α. var. *congolensis*.

Stipellae absent, leaflets usually 3 . . . . . . β. var. *bongensis*.

Leaf rhachis usually not, or hardly, prolonged beyond distal leaflet pair:

Fruiting pedicel rarely over 2 mm. long, stamens not over 7 mm. long:

Inflorescence having 2 flowers together at the very tip of the peduncle; upper leaves 1–5-foliolate; endocarp strongly spotted . . . . 23. *geminata*.

Inflorescence usually with more than 2 flowers arising one above the other along the rhachis of the inflorescence:

Leaflet margins viscid; peduncle rarely over 2 cm. long . . . .
25A. *tanganyikensis* × *nairobiensis*.

Leaflet margins not viscid; peduncle often over 2 cm. long:

Plant ± argenteous, the hairs dense, some of them softly spreading . . . .
24. *zavattarii*.

Plant green, the hairs sparse, appressed . . . . 25. *nairobiensis*.

Fruiting pedicel up to 7 mm. long; stamens up to 13 mm.; many hairs on corolla and inflorescence dark brown . . . . 26. *hedyantha*.

C.3.d.1. **I. dendroides** *Jacq.*, Coll. Bot. **2**, 357 (1789) and Ic. Pl. Rar. **3**, t. 571 (1788 or 9); L.T.A. No. 123; F.C.B. **5**, 166 t. 13.

I. multijuga Bak., F.T.A. **2**, 83 (1871); L.T.A. No. 120.

I. palustris Vatke, Oest. Bot. Zeit. **28**, 201 (1878).

I. gilletii De Wild. & T. Dur., Bull. Herb. Boiss. Ser. 2, **1**, 12 (1900); L.T.A. No. 122.

I. sesbaniifolia Chev., Bull. Soc. Bot. Fr. **58** mem. **8**, 157 (1912); L.T.A. No. 119.

I. dalabaca Chev., Expl. Bot. Afr. Occ. Fr. **1**, 172 (1920)—nomen.

I. kengeleensis De Wild., Bull. J. Bot. Et. Brux. **8**, 152 (1923); L.T.A. No. 124.

"*I. phyllanthoides* Bak." sensu Tisserant, Bull. Mus. Hist. Nat. Par. 2nd ser. **3**, 266 (1931); non Bak.

"GUINEA" (*unknown collector* ? W holotype); SENEGAL; GAMBIA; PORT GUINEA; FR. GUINEA (*Chevalier* 22359 P type of *I. dalabaca*); SIERRA LEONE; LIBERIA; IVORY COAST; *Chevalier* 22181 P holotype of *I. sesbaniifolia*; GOLD COAST; TOGO, *Büttner* 140 P; DAHOMEY; NIGERIA; CAMEROONS; B. CONGO, *Bequaert* 3951 BR holotype of *I. kengeleensis*, *Gillet* 583 BR, 737 BR, 881 BR syntypes of *I. gilletii*; OUBANGUI CHARI, *Chevalier* 6769 P; A.-E. SUDAN; UGANDA, U1–4; KENYA, K5; K7 *Taylor* s.n. BM; TANGANYIKA, T1, T3–4, T6 *Swynnerton* 33 BM, T7–8; ZANZIBAR, *Hildebrandt* 944 istoype of *I. palustris*; PORT. E. AFR., *Kirk* s.n. holotype of *I. multijuga*; NYASALAND; S. RHODESIA; N. RHODESIA; ANGOLA.

This species varies considerably; forms with shorter pods and fewer-flowered inflorescences have been distinguished as *I. kengeleensis* and *I. multijuga* and others, with unusually long inflorescences and leaves, as *I. sesbaniifolia*, but intermediates seem to be too numerous for these species to be upheld.

C.3.d.2. **I. species aff. I. amitinae** *N.E. Br.*, Kew Bull., **1925**, 158.

Port. E. Afr., from Delagoa bay to the Transvaal and Swaziland borders, *Schlechter* 12155, *Torre* 6450 LISC, 6463 LISC, *Exell, Mendonça and Wild* 522 BM.

I. amitina (type *Rogers* 14440 Carolina distr. Transvaal) is one of a difficult and confusing group of S. African species, which resemble *I. dendroides* and *I. filipes* in having long, open, for the subsection many-flowered, inflorescences, with unusually long fruiting pedicels. Some other species in this group are *I. schinzii* N.E. Br., *I. lepida* N.E. Br., *I. longipes* N.E. Br., *I. dissimilis* N.E. Br., *I. masonae* N.E. Br., *I. galpinii* N.E. Br., and *I. acutisepala* Bak.f. The whole group requires a thorough revision.

C.3.d.3. **I. filipes** *Benth. ex Harv.*, Fl. Cap. **2**, 198 (1862); L.T.A. No. 98.

S. Rhodesia; S.W. Afr., *Een* s.n. BM; Brit. Bechuanaland; Transvaal; N. Central Cape Colony; Orange Free State, *Burke* 299, *Zeyher* 472. syntypes.

C.3.d.4. **I. heudelotii** *Benth. ex Bak.*, F.T.A. **2**, 85 (1871) emend Gillett, Kew Bull. **1955**, 577 (1956).

α. var. **heudelotii.**

I. sofa Scott-Elliot, J. Linn. Soc. **30**, 76 (1894).

I. djalonica Chev. ex Hutch. & Dalz., F.W.T.A. **1**, 391 (1928); Kew Bull. **1929**, 16; non ex Bak.f., L.T.A. 147 (1926).

Port Guinea; Fr. Guinea; Sierra Leone; Nigeria; Br. Cameroons.

β. var. **fairchildii** (*Bak.f.*) *Gillett*, Kew Bull. **1955**, 577 (1956).

I. fairchildii Bak.f., J. Bot. **70**, 252 (1932).

Port Guinea; Fr. Guinea.

C.3.d.5. **I. elliotii** (*Bak.f.*) *Gillett*, Kew Bull. **1955**, 578 (1956).

I. heudelotii var. *elliotii* Bak.f., J. Bot. **41**, 261 (1903).

Senegal; Fr. Guinea; Sierra Leone; Liberia.

C.3.d.6. **I. brevicalyx** *Bak.f.*, J. Bot. **41**, 237 (1903); L.T.A. No. 70; F.C.B. **5**, 159.

"*I. pentaphylla L.*" sensu. Bak., F.T.A. **2**, 82 pro min. parte, sensu Chiov. in Pirotta Fl. Eritrea, Ann. 1st Bot. Rom. **8**, 90 (1903), non L.

"*I. sparsa Bak.*" sensu *Schweinfurth*, Bull. Herb. Boiss. **4** app. **2**, 237 (1896), sensu Chiov. in Pirotta Fl. Eritrea, Ann. 1st Bot. Rom. **8**, 414 (1908): non Bak.

B. Congo, East (fide Cronquist); Eritrea; Ethiopia, *Schimper* (1863) 1397 holotype; Uganda, U2; Kenya, K1 Marsabit *Delamere* s.n. BM, K3–5; Tanganyika, T2.

This species seems to occur at higher altitudes than other species of this subsection (900–2400 m.). Miss Hillcoat of the British Museum first observed the curious blisters on the leaflets which provide the most useful means of recognition for the species.

C.3.d.7. **I. monantha** *Bak.f.*, J. Bot. **41**, 193 (1903); L.T.A. No. 71; F.C.B. **5**, 158.

B. Congo (Katanga fide Cronquist); Tanganyika, T1, T4; Nyasaland, *Exell, Mendonça & Wild* 1025 BM; S. Rhodesia, Lomagundi distr., *Jack* in SRGH 9874, 11944; N. Rhodesia, Abercorn distr., *Nutt* s.n. holotype, *Burtt* 6250, *Mrs. Richards* 1136, 5019, 5182; Copper Belt, *Fanshawe* 944; southern prov. *Robinson* 1116.

C.3.d.8. **I. cuneata** *Bak. ex Oliv.*, J. Linn. Soc. **15,** 92 (1876); L.T.A. No. 69; non sensu F.C.B. **5,** 159.

I. latibracteata Harms, Not. Bot. Gart. Berl. **11,** 8121 (1933).

KENYA, K7; TANGANYIKA, T1, T3, T4, 4°58′S, 30°4′E, *Cameron* 8 holotype, T8 Kilwa. distr. *Schlieben* 2458 BM isotype of *I. latibracteata.*

Older plants such as *Cameron* 8 in which the earlier, multijugate, leaves have been shed appear very different from younger plants, such as *Schlieben* 2458 in which such multijugate leaves predominate. E. G. Baker's statements, J. Bot. **41,** 193 and L.T.A. 114 that the type locality is in "British Central Africa" (N. Rhodesia) are erroneous. No material has been seen from N. Rhodesia. The type locality is given in detail in Oliver's original paper.

C.3.d.9. **I. manyoniensis** *Bak.f.*, J. Bot. **71,** 340 (1933).

TANGANYIKA, T5, Manyoni distr., *B. D. Burtt* 3643 BM holo K isotype.

Burtt 4656 distributed as this species seems to be a form of *I. tanganyikensis* Bak.f.

C.3.d.10. **I. micropetala** *Bak.f.*, J. Bot. **71,** 340 (1933).

TANGANYIKA, T5, Manyoni distr., *B. D. Burtt* 3608 BM holo K isotype.

C.3.d.11. **I. tanganyikensis** *Bak.f.*, Kew Bull. **1932,** 96.

α. var. **tanganyikensis.**

I. cuneata sensu Cronquist, F.C.B. **5,** 159 non Bak. ex Oliv.

B. CONGO (east), *Germain* 1145; ETHIOPIA, Harar Prov., *Bally* 10078; UGANDA, U2–4; KENYA, K3–5; TANGANYIKA, T1–4, T5 *B. D. Burtt* 1738 BM holotype, T6.

β. var. **paucijuga** *Gillett* var. nov.

A var. *tanganyikense* foliolis 3–5, non 7–15, saepe pendulis, differt.

KENYA, K7 between Samburu and MacKinnon Rd. *c.*350 m., side of road through *Acacia-Commiphora* scrub; perennial much branched subshrub, leaves sticky glandular, standard deep pink, wings paler pink, keel greenish pink, 30.8.1953, *Drummond & Hemsley* 4056 holotype. Steppe near Bura hills, *c.*1000 m., small bright purple red flowers, 16.9.1901, *C. Uhlig* s.n. EA.

Intermediates, perhaps of hybrid origin, occur between *I. tanganyikensis* and *I. vohemarensis* on the one hand and *I. nairobiensis* on the other.

C.3.d.12. **I. mwanzae** *Gillett* sp. nov.

Ab. *I. tanganyikense* habitu annua, fructibus subcomplanatis, seminibus ad terminos corrugatis differt.

Herba annua *c.*20 cm. alta, ramis patentibus, pilis sparsis biramosis patentibus, ad 1·5 mm. longis, vestitis. *Folia* 3–7-foliolata; stipulae lineares, 2–4 mm. longae, parce pilosae; rhachis complanata, superne canaliculata, parce pilosa, in tota 7–12 mm. longa, petiolo 2–3 mm. longo incluso, ultra foliola lateralia *c.*2 mm. prolongata; stipellae nullae; foliola 0·5 mm. petiolulata, anguste elliptico-oblonga, subtus strigis *c.*1 mm. longis sparse aspersa, superne parce puberula, apice mucronata, marginibus paulo incrassatis, ut videtur iuventute, ut mucrone apicale, glandulosis, terminale lateralibus *c.*7 mm. longis 2 mm. latis paulo majore. *Racemi* 2-flori, axe tenui, flavo, glabro, 11–18 mm. longo, pedunculo 9–15 mm. longo incluso; bracteae mox deciduae, haud visae; pedicelli pilosi, 1–1·3 mm. longi. *Calyx* pilosus, nervis exceptis, translucens, *c.*1·8 mm. longus, ad mediam partitus, lobis aequalibus basin deltoideis, superne subulatis, apice glanduloso incrassatis. *Vexillum* extus puberulum, late ellipticum, *c.*3·5 mm. longum 2·5 mm. latum. *Alae* dolabriformes, glabrae, *c.*3 mm. longae. *Carina* extus pubescens, calcare *c.*0·4 mm. longo, haud rostrata, *c*.3·4 mm. longa. *Stamina* omnia fertilia, filamentis *c*.4 mm. longis, antheris basin muticis, suborbicularibus, *c*.0·6 mm. longis, mucrone apicale rubro *c*.0·15 mm. longo incluso. *Ovarium* pauce strigosum, *c*.2·5 mm. longum, *c*.12-ovulatum, stylo *c*.1 mm. longo, paulo infra stigma globosum valde curvato. *Legumen* fere glabrum, ad suturam strigis paucis appressis obsitum, subcomplanatum, 14–16 mm. longum, *c*.1·5 mm. latum, *c*.0·9 mm. crassum, 11–12-spermum, endocarpio maculato. *Semina* brunnea, aspectu laterale quadrata, foveolata, aspectu terminale elliptica longitudinale canaliculata.

Tanganyika, T1, Buhungukira area Kwimba-Mwanza districts, *c*.32°50′E, 3°12′S, common in pastures, upright, flowers red, 4.5.1932, *V. V. Rounce* 21 K iso, EA holotype.

C.3.d.13. **I. elwakensis** *Gillett* sp. nov.

Ab *I. tanganyikense* racemis longioribus, 3–15-floris, foliolisque latioribus differt.

Herba perennis, radice lignoso, ramis ± procumbentibus, pilis medifixis ± patentibus haud dense vestitis. *Folia* 5–9-foliolata; stipulae 1–2 mm. longae ± subulatae; rhachis sparse subpatente strigosa, 13–18 mm. longa, petiolo 2–3 mm. longo incluso, ultra foliola lateralia *c*.2 mm. prolongata; foliola subaequalia, late elliptica vel obovata, marginibus apicibusque incrassatis glandulosis, *c*.5 mm. longa, 3 mm. lata, superne strigis *c*.0·5 mm. longis, inferne strigis *c*.1 mm. longis obsita. *Racemi* 3–15-flori, axe tenui fere glabro 25–50 mm. longo, pedunculo 8–20 mm. longo incluso; bracteae ad 1 mm. longae mox deciduae; pedicelli hirsuti *c*.1 mm. longi. *Calyx* strigosus, tubo, praeter nervos, translucente, 1 mm. longo, lobis basin deltoideis superne subulatis 0·8 mm. longis. *Vexillum* extus pubescens, suborbiculare, 4·5 mm. longum, 4 mm. latum. *Alae* glabrae, dolabriformes, 4·5 mm. longae 2 mm. latae. *Carina* margine et ad commissuram pilosa calcare 0·6 mm. longo, haud rostrata, 5 mm. longa. *Stamina* omnia fertilia, filamentis fere 5 mm. longis, antheris ovalibus, basin muticis, *c*.0·5 mm. longis, mucrone apicale 0·1 mm. longo incluso. *Ovarium* fere glabrum, *c*.3·5 mm. longum, *c*.10-ovulatum, stylo 1·5 mm. longo fere ad basin curvato, stigmate minuto. *Legumen* fere glabrum vel sparse strigosum, 12–15 mm. longum, 1·5–2 mm. latum, *c*.8-spermum, endocarpio maculato. *Semina* matura haud visa.

Kenya, K1, 13 km. S of El Wak on road to Wajir, 40°55′E, 2°40′N, *c*.400 m., in rich *Commiphora-Acacia* open scrub on red sandy loam, flowers brick red, 26.5.1952, *Gillett* 13347 K holo. EA isotype. Wajir, 40°4′E, 1°45′N, 240 m., rainfall 200 mm., red sandy soil, *Commiphora-Acacia* scrub with much *Delonix elata*, overgrazed, 27.5.1952, *Gillett* 13370.

C.3.d.14. **I. ambelacensis** *Schweinf.*, Bull. Herb. Boiss. **4** app. **2**, 237 (1896); L.T.A. No. 72; F.C.B. **5**, 157.

I. pauciflora De Wild., Bull. J. Bot. Et Brux. **8**, 153 (1923): non Eckl. & Zeyh.

I. wildemani Bak.f., L.T.A. 126, No. 111 (1926).

I. conradsii Bak.f., J. Bot. **70**, 252 (1932).

"*I. sp. an nov.?*" Bak.f., J. Bot. **41**, 238 (1903).

B. Congo, East *Bequaert* 5480 BR holotype of *I. pauciflora* De Wild and *I. wildemani*; Eritrea, Bogos, *Hildebrandt* 587 BM, Ambelaco (*Schweinfurth* (1894) 293 holotype); A.-E. Sudan, *A. W. Peers* T.018; Uganda, U1–4; Kenya, K4–5; Tanganyika, T1 Ukerewe island, *Conrads* 165 (B† holo.) BM fragment isotype of *I. conradsii, Conrads* in EA 10,366, T3, T6.

The type has not been seen; *Hildebrandt* 587, from a nearby locality, agrees very well with the description, and might be taken as a neotype, should *Schweinfurth* 293 prove to have been wholly destroyed.

C.3.d.15. **I. kongwaensis** *Gillett* sp. nov.

Ab *I. ambelacense* petiolo longiore, legumine breviore, foliolis folium superiorum paucioribus differt.

Herba annua *c.*30 cm. alta ramis diffusis pilis patentibus sparse vestitis. *Folia* primaria 7–9-foliolata, posteriora 3–5-foliolata; stipulae filiformes pilosae, ad 4 mm. longae; rhachis patente strigosa ad 19 mm. longa, petiolo *c.*5 mm. longo incluso, ultra foliola lateralia *c.*1 mm. prolongata; foliola cuneato-oblonga, apice rotundata, ad 12 mm. longa, 3 mm. lata, superne breviter, inferne longe strigosa. *Racemi* 4–5-flori; axis pubescens, 15–27 mm. longus, pedunculo 11–20 mm. longo incluso; bracteae ut videtur nullae;pedicelli pubescentes, ad 1 mm. longi. *Calyx* hirsutus, nervis exceptis, translucens, tubo 0·5–1 mm. longo, lobis basin deltoideis, superne subulatis, fere 2 mm. longis. *Vexillum* extus pubescens, ovatum, apice acutum, 3·8 mm. longum, 3 mm. latum. *Alae* glabrae, dolabriformes, 3 mm. longae, 1 mm. latae. *Carina* extus pubescens, calcare 0·5 mm. longo, 0·1 mm. rostrata, 4 mm. longa. *Stamina* omnia fertilia, filamentis 3·5–4 mm. longis, antheris basin muticis, oblongis, 0·5 mm. longis, mucrone apicale 0·1 mm. longo incluso. *Ovarium* pilosum, *c.*1·5 mm. longum, 5–6-ovulatum, stylo inferne piloso superne glabro, 3 mm. longo, 1 mm. infra stigma capitatum curvato. *Legumen* pubescens, 4–6-spermum, 7–9 mm. longum, rostro 1–1·5 mm. longo (styli base persistente) haud incluso, 1·3 mm. latum, 0·8 mm. crassum, endocarpio obscure maculato. *Semina* aspectu laterale subquadrata, aspectu terminale anguste elliptica.

Tanganyika, T5, Kongwa, 1000 m., weed of agricultural land cleared from *Commiphora* bush, prostrate annual, corolla red, 19.5.1950, *B. Anderson* EA holo., K isotype; ibid. in field A4, N.E. corner, herb to 30 cm. tall. spreading branches, corolla red darker towards centre, 1.3.1950, *Holly* 9; ibid. pasture, red soil, 12.4.54, *Brzostowski* in EA 10848 EA.

C.3.d.16. **I. wituensis** *Bak.f.*, J. Bot. **41**, 243 (1903); L.T.A. No. 97.

α. var. **wituensis.**

Kenya, K7, Witu, *Thomas* 207 holotype; Tanganyika, T3, T6 *Vaughan* 2841 BM; Port. E. Afr., Cabo Delyado, nr. Macomia, 40°E, 12°10′S, *Barbosa in Mendonça* 2070 LISC.

β. var. **occidentalis** *Gillett,* Kew Bull. **1955**, 578 (1956).

N. Nigeria, *Lely* P693 holotype.

C.3.d.17. **I. mendonçae** *Gillett* sp. nov.

I. wituensi affinis; ab ea, radice perennante sublignosa, racemis pluri- (ad 15-) floris, foliolis angustioribus, longioribus, superne sparse strigosis, differt.

Herba erecta ad 60 cm. alta, radice sublignosa, perennante, ramis flavo-viridibus, glabris. *Folia* 5–9-foliolata; stipulae glabrae, lanceolatae, *c.*2 mm. longae; petiolus glaber, subnullus vel ad 1 mm. longus; rhachis glabra, ad 2 cm. longa, ultra folia lateralia vel breviter vel ad 5 mm. prolongata; stipellae brevissimae; petioluli flavi, glabri, *c.*0·7 mm. longi; foliola anguste oblonga, apice emarginata, truncata, rotundata vel late acuta, inferne fere glabra, superne pilis medifixis appressis *c.*0·5 mm. longis sparse obsita, margine, ut in *I. tanganyikensi* et *I. wituensi*, glanduloso incrassata, ad 10 [13] mm. longa et 2 mm. lata. *Racemi* axillares, glabri, 4–15-flori, pedunculo ad 15 mm., rhachide florifera ad 4 cm. longa; bracteae lineares, mox caducae, ad 1 mm. longae; pedicelli 1 mm. longi, post anthesin erecti, ad 2 mm. longi. *Calyx* pilis medifixis appressis sparse obsitus, tubo 0·7–1 mm. longo, lobis deltoideis, 2 superioribus *c.*0·7 mm., 3 inferioribus 1 mm. longis. *Vexillum* fere glabrum, late orbiculare, *c.*4 mm. longum, 5 mm. latum. *Alae* glabrae, 4 mm. longae, unguiculo 0·2 mm. longo incluso, 1·7 mm. latae. *Carina* haud rostrata, margine pilosa, lateraliter 0·8 mm. calcarata, in tota 5 mm. longa. *Stamina* omnia fertilia, filamentis 3·8–4·5 mm. longis, antheris ovatis, basin muticis, fuscis, 0·7 mm. longis, mucrone apicale 0·2 mm. longo incluso. *Ovarium* glabrum, fere 4 mm. longum; stylus *c*.1 mm. longus, glaber, basin per angulum 60° curvatus, stigmate capitato 0·2 mm. diametro. *Legumen* glabrum, *c*.10-spermum apice 1 mm. rostratum, subcylindricum 14–16 mm. longum, 2 mm. latum, 1·5 mm. crassum, endocarpio maculato. *Semina* brunnea, subcuboidea, ± foveolata, 1·2 mm. longa, 1 mm. lata. 0·7 mm. crassa.

PORT. E. AFR., Inhambane distr. between Guissico and Chicoma, *c*.34°30′E, 24°40′S, on sandy soil liable to floods, perennial viscid herb, flowers with red wings and keel, 7.12.1944, *F. A. Mendonça* 3315 LISC holo. K isotype. Chibuto distr., Porto do Alto Changane, grassy fields bordering the river Changane, woody herb, flowers red, 14.7.1944, *A. R. Torre* 6748 LISC. Same locality, alt. 35 m., herb 70 cm. tall with sticky leaves and small pink flowers, 31.10.1935, *A. Lea* 138 K, (PRE).

C.3.d.18. **I. sparsa** *Bak.*, F.T.A. **2**, 79 (1871); L.T.A. No. 74, non sensu Schweinfurth, Bull. Herb. Boiss. **4** app. 2, 237 (1896) nec Chiov. in Pirotta Fl. Eritrea Ann. Ist Bot. Rom. **8**, 414 (1908) nec Cronquist, F.C.B. **5**, 158.

N. ETHIOPIA, *Schimper* (1863–8) 204 K holo. BM isotype, (1852) 2009 BM.

C.3.d.19. **I. tenuis** *Milne-Redhead*, Kew Bull. **1936**, 472.

α. ssp **tenuis.**

UGANDA, U1 Karamoja, *Thomas* 3388; KENYA, K2; TANGANYIKA, T1 *Bax* 151 holotype; N. RHODESIA, Abercorn distr., *Mrs. Richards* 639, 4772.

β. ssp **major** *Gillett*, var. nov.

A ssp *tenue* leguminibus majoribus (*c*.20, non infra 15 mm. longis), foliolis majoribus (saepe 15 mm. excedentibus), differt.

S. RHODESIA, Lomagundi distr. Trelawney, 30°28′E, 17°30′S, 1200 m. tobacco research station, 29.3.1944, *R. W. Jack* in SRGH 11938 K holotype, 9584 SRGH, 9610 SRGH, 9835 SRGH, 9838 SRGH. Hartley district, Msengezi, Granite sandveldt—cleared bush, April 1955, *A. P. Conradie* 6.

C.3.d.20. **I. dissitiflora** *Bak. ex Oliv.*, J.L. Soc. **15**, 93 (1876); L.T.A. No. 105.

TANGANYIKA, T4 *Cameron* 41 holotype, *H.M. Lloyd* 71.

C.3.d.21. **I. vohemarensis** *Baill.*, Bull. Soc. Linn. Par. **1**, 400 (1883).

I. viscosa Lam. var. *subglabra* A. Rich., Tent. Fl. Ab. **1**, 181 (1846).

"*I. pentaphylla L.*" sensu Bak., F.T.A. **2**, 89 (1871) pro. maj. parte, non Murr. in L.

"*I. suaveolens J. & S.*" sensu Taub. in Engl. Pflwelt. O. Afr. C, 210 (1895); Bak.f., J. Bot **41**, 240 (1903); L.T.A. No. 110 pro maj. parte; sensu Brenan, Check List Tanganyika trees and shrubs 427; Cronquist, F.C.B. **5**, 160; non Jaub & Spach.

Anila pentaphylla (L.) (per errore "Burch.") O. Ktze var. *trichocarpa* O. Ktze, Rev. Gen. **3**: **2**, 52 (1898).

I. uhehensis Harms, Eng. Bot. Jahrb. **28**, 402 (1900); L.T.A. No. 108.

I. subquadriflora Hochst. ex Chiov. in Pirotta Fl. Eritrea, Ann. 1st. Bot. Rom. **8**, 90 (1903) (nomen).

I. suaveolens Jaub. & Spach var. *subquadriflora* Chiov. in Pirotta Fl. Eritrea, Ann. 1st Bot. Rom. **8**, 414 (1908).

I. minimifolia Chiov., Racc. Bot. Miss. Consol. Kenya 29 (1935). ?

I. suaveolens var. *pseudocongolensis* Chiov. in Cufodontis, Miss. Biol. Boran. 75–77, fig. 14 (1939).

B. Congo, East (fide *Cronquist*); Ethiopia, Chire *Quartin Dillon* s.n. (P holo.) K isotype of *I. viscosa* var. *subglabra*, *Schimper* (1852) 2001 BM isotype of *I. subquadriflora* and *I. suaveolens* var. *subquadriflora*, *Cufodontis* 548 Fi holotype of *I. suaveolens* var. *pseudocongolensis*; Uganda, U1, U3–4; Kenya, K1, K3, K4 (Nyeri 29.7.1929, *Balbo* 598 MOD †; Imenti near Meru, 1200 m., 5.4.1929, *Balbo* 866 MOD † syntypes of *I. minimifolia*), etc., K6–7; Tanganyika, T1–3, T5–6, T7 *Goetze* 508 (408) isotype of *I. uhehensis*, 8; Zanzibar; Pemba; Port. e. Afr., Mozambique, 5.4.1894, *O. Kuntze* s.n. isotype of *A. pentaphylla* var. *trichocarpa*; Madagascar, *Pervillé* s.n. isosyntype.

The syntypes of *I. minimifolia* were, almost certainly, destroyed in the war. Chiovenda's description agrees well with certain states of *I. vohemarensis* in which the leaves are stunted but, in view of the difficulty of this subsection of the genus, it is desirable that the neighbourhood of Nyeri and of Imenti 37°42′, 0°5′S should be investigated.

In F.T.A. *I. vohemarensis* was confused with the following species:

I. glabra *L.*, Sp. Pl. 751 (1753); Fl. Br. Ind. **2**, 95.

I. pentaphylla Murr. in L. Syst. Veg. ed. **13**, 564 (1774); non Burch. ex Harv., Fl. Cap. 2, 178; nec sensu Bak., F.T.A. **2**, 82.

   I. fragrans Retz., Obs. 4, 29 (1786).

This species seems to be confined to India and Ceylon. Of the African specimens cited in F.T.A. 2, 82 the majority are *I. vohemarensis* Baill. while *Schimper* (1853) 531 is *I. brevicalyx*. The true Indian plant has long, stiff, spreading, sub-simple hairs, the young leaflets and calyx lobes usually gland-tipped, large lanceolate stipules and bracts, and glabrous fruits with unusually broad sutures. It is more closely related to *I. pilosa* than to any species of §§**Dissitiflorae** as here defined. Two specimens from Madagascar, *Baron* 4810 and *Bojer* s.n. identified by Drake, Hist. Pl. Madag. 156 as *I. pentaphylla* are not that species, but a distinct, apparently undescribed, species in §§**Dissitiflorae** remarkable for its very long petioles.

C.3.d.22. I. congolensis *De Wild. & Th. Dur.*, Bull. Herb. Boiss. 2nd ser. **1**, 11 (1901); L.T.A. No. 114.

α. var. **congolensis.**

"*I. geminata Bak.*" sensu Hutch. & Dalz., F.W.T.A. **1**, 391 (1928) p.p.; non Bak.

"*I. sparsa Bak.*" sensu Cronquist, F.C.B. **5**, 158, non Bak.

I. adami Berhaut, Fl. Seneg. 52 (1954) sine latine, Bull. Soc. Bot. Fr. **101**, 374 (1955) (e descr.)

Senegal, Galam *Heudelot* 265 P (Kontango-Koular *Adam* 10148 P holotype of *I. adami*); Gambia, *Hayes* 560; Fr. Sudan, *Chevalier* 3219; Fr. Niger Col., *Hagerup* 499 BM; N. Nigeria, *Lely* 539, 629, 646, *Dalziel* 20, *Batten Poole* in FHI 13177 FHI, *Keay* in FHI 20144, *Ross* 265 FHI; B. Congo, *J. Gillet* 732 BR holotype, Chari, *Chevalier* 5553 P; Oubangui, *Le Testu* 3094; Uganda, U3, *Chandler* 901.

β. var. **bongensis** (*Bak.f.*) *Gillett* stat. nov.

I. sparsa Bak. var. *bongensis* Bak.f., J. Bot. **41**, 237 (1903).

A.-E. Sudan, *Schweinfurth* 2526 K iso., BM holotype.

This variety is in some ways intermediate between *I. sparsa* and *I. congolensis*. It agrees with *congolensis* in characters 1, 3 and 4 listed below, with *sparsa* in No. 2. The inflorescences are 2-flowered. The leaflets are narrower than in either species. Tisserant, Bull. Mus. Hist. Nat. Par. 2nd ser. **3**, 263 (1931) lists several specimens under this variety. Of these I have seen two, *Chevalier* 3219 and *Heudelot* 265 which are both var. *congolensis*.

Dr. Cronquist treats *I. congolensis* as conspecific with *I. sparsa*. I find the following differences between the two species and consider it preferable to retain them as distinct, at least until much more is known.

<table>
<tr><td>I. sparsa</td><td>I. congolensis</td></tr>
<tr><td>1. Petiole often very short (under 1 mm.).</td><td>Petiole not less than 2 mm.</td></tr>
<tr><td>2. Stipellae absent, or nearly so.</td><td>Stipellae well developed.</td></tr>
<tr><td>3. Fruit 9–11 mm. long.</td><td>Fruit 13 mm. or more long.</td></tr>
<tr><td>4. Leaflets appressed pilose above.</td><td>Leaflets glabrous or glabrescent above.</td></tr>
<tr><td>5. Flowers 1–2.</td><td>Flowers 2–4.</td></tr>
</table>

In West Africa *I. congolensis* has been much confused with *I. geminata*; the following differences separate them.

<table>
<tr><td>I. geminata</td><td>I. congolensis</td></tr>
<tr><td>1. Leaves subtending upper inflorescences usually 3-foliolate.</td><td>Leaves subtending all inflorescences 5–7, very rarely 3-foliolate.</td></tr>
<tr><td>2. Leaf rhachis often not prolonged beyond lateral leaflets.</td><td>Leaf rhachis always clearly prolonged beyond lateral leaflets.</td></tr>
<tr><td>3. Upper leaf surface appressed strigose.</td><td>Upper leaf surface glabrous or glabrescent.</td></tr>
<tr><td>4. Leaflets oblanceolate with a long tapering base.</td><td>Leaflets oblong-obovate ± abruptly narrowed at the base.</td></tr>
<tr><td>5. Flowers 2, their pedicels inserted together at the tip of the peduncle.</td><td>Flowers 2–4, their pedicels inserted one above the other on the inflorescence axis.</td></tr>
<tr><td>6. Endocarp distinctly spotted.</td><td>Endocarp obscurely spotted.</td></tr>
</table>

In addition the filaments in West African specimens of *congolensis* seem to be consistently shorter than those of *geminata* (3–4 mm. instead of 4–5 mm.) but this is not a valid character for *congolensis* as a whole, as there seems to be a cline from a filament length of *c.*3 mm. in Senegal to 5 mm. in Uganda.

C.3.d.23. I. geminata *Bak.*, F.T.A. **2**, 81 (1871); L.T.A. No. 93.

"*I. sp. aff. geminata Jacq.*" sensu Chevalier, Expl. Bot. Afr. Occ. Fr. 173.

FR. SUDAN, *Chevalier* 24729 P, 24772, 24812 P, 24815 P; DAHOMEY, *Chevalier* 22694 P; NIGERIA, *Dalziel* 1221, *Rowland* s.n., *Barter* 966 holotype, *Barter* s.n.

Riva 77 from Dscharibale in S. Ethiopia, 38°35′E, 5°35′N., which was identified by Harms as *I. geminata* has not been seen. This record requires confirmation.

C.3.d.24. I. zavattarii *Chiov.* in Cufodontis. Miss. Biol. Boran. 77, fig. 16 (1939).

S. ETHIOPIA, Javello, *Cufodontis* 467 F1 holotype, *Mooney* 5509; KENYA, K1, S. Turkana Kailongol, 1800 m., *Champion* T301.

C.3.d.25. I. nairobiensis *Bak.f.*, Fed. Rep. **10,** 124 (1911); L.T.A. No. 141, including var. *angusta* Bak.f., L.T.A. 133 (1926).

KENYA, K3 *Battiscombe* 728 holotype of var. *angusta*, K4 (*Winkler* 4166 BRSL holotype); TANGANYIKA, T2.

C.3.d.25A. I. tanganyikensis × nairobiensis?

KENYA, K3 Kipkarren—Soy Rd., 1500–1800 m., Aug. 1931, *Mrs. Brodhurst-Hill* 1925 K, EA, 212 K, EA. Uasin Gishu, 2160 mm., Jan. 1932. *Miss Harvey*

3481 K, EA. Turbo-Eldoret 1950 m., 5.9.1952, *Bogdan* 3598 K, EA. Eldoret distr., 2070 m., 13.4.1951, *Williams* G119. Thompson's falls, 2310 m., 24.8.1950, *Miss Lacey* 44B. Ol jororok, 2100 m., *R. Pierce* 2698 K, EA.

These plants are intermediate between *I. tanganyikensis* and *I. nairobiensis* in the way shown below, possibly as a result of hybridisation.

	I. tanganyikensis	*? hybrids*	*I. nairobiensis*
1. Leaf rhachis prolonged beyond lateral leaflets	Yes	Usually not	No
2. Margins of leaflets	Viscid	Viscid	Not viscid
3. No. of flowers in inflorescence	2–4	3–7	4–12
4. Length of peduncle	0·6–1·4 cm.	1·2–2 cm.	1·5–6 cm.
5. Indumentum of pods	Usually nil	Sparse appressed	Sparse appressed

C.3.d.26. **I. hedyantha** *Eckl. & Zeyh.*, Enum. Pl. Afr. Austr. 243 (1836); L.T.A. No. 140.

I. goetzei Harms, Eng. Bot. Jahrb. **28**, 402 (1900); L.T.A. No. 142.

I. inyangana N. E. Brown, Kew Bull. **1906,** 102; L.T.A. No. 139.

KENYA, K4; TANGANYIKA, T5, T7, *Goetze* 628 (B†) BM photo. holotype of *I. goetzei*; PORT. E. AFR., Lebombo mts.; NYASALAND; N. RHODESIA, Abercorn distr.; S. RHODESIA, *Cecil* 174 holotype of *I. inyangana*; TRANSVAAL; NATAL; E. CAPE PROV., *Ecklon & Zeyher* 1614 isotype.

Subsection C.3.e. **Spinosae** (*Bak.*) *Gillett* stat. nov.

"group" *Spinosae* Bak., F.T.A. **2**, 67 (1871); L.T.A. 113 p.p.

Lectotype species *I. spinosa* Forsk.

Centre of distribution N.E. Trop. Africa.

Suffruticose plants of arid areas. Leaflets 3–9, the lateral ones opposite. Inflorescence few-flowered, its axis becoming a spine. Fruits subcylindrical, erect or spreading; not sharply deflexed. Endocarp spotted.

As here delimited this seems to be a natural group, related to subsect. **Dissitiflorae.** Superficially similar suffruticose species with spinose inflorescences have, however, been evolved in arid areas in several other parts of the genus; e.g. *I. circinnata* Benth. ex Harv. and *I. sessilifolia* DC. in subgen. **Amecarpus,** *I. spiniflora* Hochst. & Steud ex Boiss, *I. trita* L.f. and *I. pungens* E. Mey in subsect. **Tinctoriae** and *I. intricata* in subsect. **Alternifoliolae.**

Axis of inflorescence ending in a glabrescent spine even in the bud stage:
    Pods covered with stiff glandular hairs; leaves up to 4-jugate, leaflets 10-17 mm. long . . . . 1. *rothii.*

    Pods free from glandular hairs; leaves 1-, rarely 2-or 3-jugate, leaflets rarely over 10 mm. long:
        2–5 flowers borne on the upper half of the inflorescence axis; flowering pedicels up to 1 mm. long; petiole often 2 mm. or more long   2. *spinosa.*

        1–2 flowers borne at base of inflorescence axis; flowering pedicels up to 4 mm. long; petiole rarely exceeding 1 mm. . . . . 3. *basiflora.*

Axis of flowering inflorescence argenteous, not spinose, ending in a group of buds; the fruiting inflorescence alone spinose . . . . 4. *longispina.*

C.3.e.1. **I. rothii** *Bak.*, F.T.A. **2**, 77 (1871); L.T.A. No. 67.

ETHIOPIA, *Roth* s.n. holotype, *Gillett* 5230.

50

C.3.e.2. **I. spinosa** *Forsk.*, Fl. Aeg. Arab. 137 (1775); Jaub. & Spach., Ill. Pl. Or., t. 490–1; Bak., F.T.A. **2**, 77 pro maj. parte (synonyma exclusa); Bak.f., L.T.A. No. 66, pro maj. parte (var. *spiniflora* exclusa) including var. *microphylla* A. Rich., Tent. Fl. Ab. **1**, 185 (1846).

"*I. intricata Boiss.*" sensu Hutch. & Bruce, Kew Bull. **1941**, 118 pp.; non Boiss.

Hejaz; Yemen, *Forskal* s.n. C holotype; Aden, Hadhramaut; A.-E. Sudan; Eritrea; Ethiopia, *Quartin Dillon* s.n. isotype of var. *microphylla*; Brit. Somaliland, *Gillett* 3951; Kenya, K1–2, K4, K6–7; Tanganyika, T3.

In the area from Brit. Somaliland to the Voi district of Kenya, besides the ordinary form, plants occur in which the leaves are largely 5- and occasionally 7-foliolate. Elsewhere 3-foliolate leaves only have been seen. *Gillett* 4486 from Brit. Somaliland is abnormal in having numerous flowers spread along nearly the whole length of the inflorescence. This could be a result of introgressive hybridisation with *I. spiniflora* q.v. but otherwise its characters are those of *I. spinosa*.

C.4.e.3. **I. basiflora** *Gillett* sp. nov.

Ab *I. spinosa* ut in clavi distincta.

Suffrutex spinosus *c*.50 cm. altus, ramis rigidis subargenteis vel viridibus, pilis apice ± patentibus dense vestitis. *Folia* saepius digitatim trifoliolata, nonnunquam rhachide ad *c*.0·5 mm. ultra foliola lateralia prolongata, rarius unifoliolata; petiolus pilis ± patentibus argenteus, saepius 0·5–1 mm., rarius ad 2 mm. longus; stipulae subulatae 0·5 mm. longae; foliola saepius complicata, oblongo-cuneata, apice truncato-rotundata, paulo recurvata, subtus dense, superne sparse, strigosa, strigis *c*.0·5 mm. longis, *c*.5 mm. longa, 2 mm. lata. *Racemus* 2-florus, pedicellis pilosis ad 4 mm. longis, ad basin axis spiniformis subglabri, straminei, ad 15 mm. longi, positis. *Calyx* pilosus ad 1·7 mm. longus, lobis aequalibus, deltoideis, *c*.0·6 mm. longis, inclusis. *Vexillum* extus strigosum, oblongo-ellipticum, *c*.6 mm. longum, 3 mm. latum. *Alae* glabrae, ± oblongae, *c*.5 mm. longae, unguiculo 0·3 mm. longo incluso. *Carina* extus pubescens, calcare 0·6 mm. longo, haud rostrata *c*.6 mm. longa. *Stamina* omnia fertilia, filamentis *c*.5·5 mm. longis, antheris late ellipticis, basin muticis, *c*.0·7 mm. longis, mucrone apicale *c*.0·1 mm. longo incluso. *Ovarium* pilosum *c*.5 mm. longum, *c*.10-ovulatum, stylo glabro, *c*.2·5 mm. longo, paulo infra mediam curvato, stigmate ovale. *Legumen* patens (vel suberectum vel subreflexum) sparse strigosum, strigis *c*.0·4 mm. longis, subquadrangulare, ad 21 mm. longum, *c*.2·2 mm. latum, 2 mm. crassum, 7–9-spermum, endocarpio maculato. *Semina* suboblonga.

Tanganyika, **T2**, Masai distr. Ngorongoro crater, near Siedentopf's, in dry grassland, 1500 m., 9.4.1941, *Bally* 2272; ibid. nr. Laroda, 1770–1780 m., dwarf shrub 25 cm. tall. 23.7.1926, *Peter* 43143. Arusha distr. Oldoinyo Sambu *c*.1350 m., roadside, bushy, 50 cm. tall, flowers pink, common gregariously, 31.1.1932, *St. Cl. Thompson* 270 holotype; Ibid 1800 m. dwarf shrub 50 cm. tall, bad weed on grazing land, Feb. 1926, *Haarer* B152; ibid, 1500 m., dry stony grassland with *Aloe* and stunted *Commiphora* sp., much branched shrublet up to 70 cm. tall, *G. W. Ivens* 200 EA. Engare Nanyuki, small shrub *c*.35 cm. tall, flowers red, abundant in heavily grazed *Themeda triandra* veldt, 11.10.1945, *Van Rensberg* 188 EA. Ibid. roadside, flowers bright pink, Oct. 1933, *Mainwaring* in *Napier* 3043. Near Ol Kokola, *c*.36°36′E, 3°15′S, 1200–1800 m., closely grazed *Themeda* grassland without trees on volcanic ash, in weedy roadsides, shrublet –60 cm., branches spreading, some decumbent at base, flowers bright pink, occasional, 7 and 19.4.1955, *P. Huxley* 194, 200. Meru, "common throughout west Masailand" *Hornby* H40/32/15 K, EA. Moshi distr., Engare Nairobi, 1500 m., 23.8.1936, *Davies* 1197 K, EA. Ibid, 1380 m., low bush 25–35 cm., flowers pink, 13.11.1930, *Sanders* 5 K, EA. Ibid. 1360 m., on thin rocky soil, 32 cm. tall. 10.3.1933, *Milne* 15 K, EA. **T3** Pare-Lushoto districts, between Hedaru and Ukomazi, 450–600 m., 13.6.1915, *Peter* K178; 24.7.1915, *Peter* K205; 7.6.1926, *Peter* 41024.

It may be noted (1) that the range of this species, though adjacent to, does not overlap, that of *I. spinosa*, (2) that, while *I. basiflora* has recently been repeatedly described as common, it was apparently not collected before 1915, although several early collectors traversed its area. Being an unpalatable plant it has probably become much more common as a result of the increase in the intensity of grazing caused by the establishment of law and order, and the reduction in the number of lions.

C.3.e.4. **I. longispina** *Bak.f. ex Gillett* sp. nov.

"*I. spinosa Forsk.*" sensu Bolus, Bolus and Glover, Ann. Bol. Herb. **1**, 14; non Forsk.

Ab *I. spinosa* racemo florifero brevi, molle, haud spinoso; racemoque fructifero spinoso longiore differt.

Suffrutex c.30 cm. altus, ramis junioribus rigidis argenteis pilis appressis densissime vestitis, vetustioribus calvescentibus brunneis. *Folia* argentea, ubique pilis appressis dense vestita, 3–7-foliolata; stipulae deltoideae, c.0·5 mm. longae; rhachis, ultra foliola lateralia haud prolongata, ad 4 mm. longa, petiolo 1–2 mm. longo incluso; foliola cuneato-obovata, apice obtusa, paulo recurvata, c.2 mm. longa, 1·5 mm. lata. *Racemus* 5–8-florus ad anthesin c.7 mm. longus, pedunculo 1–3 mm. longo incluso, ubique argenteus, bracteis c.0·6 mm. longis, mox deciduis, pedicellis c.0·9 mm. longis; post anthesin ad 30 mm. elongatus, induratus, spiniformis, calvescens, flavus, pedicellis suberectis vel patentibus *Calyx* argenteus, c.2 mm. longus, lobis subaequalibus deltoideis, c.1 mm. longis, inclusis. *Vexillum* extus pubescens suborbiculare, c.4 mm. longum et latum. *Alae* glabrae c.3·8 mm. longae, unguiculo 0·2 mm. longo incluso. *Carina* ad commisuram pilosa, calcare 0·6 mm. longo, haud rostrata, c.4 mm. longa. *Stamina* omnia fertilia, filamentis c.4 mm. longis, antheris ellipticis basin muticis 0·7–0·9 mm. longis, mucrone apicale 0·1–0·2 mm. longo incluso. *Ovarium* c.3 mm. longum, strigosum, c.10-ovulatum, stylo 1·2 mm. longo, basin curvato, stigmate globoso. *Legumen* cylindricum sparse strigosum.

S.W. AFRICA, Eastern Karasberg, sandy plains N.E. of Narudas süd, 1320 m., bush 30 cm. tall, corolla red purple, 28.12.1912, *H. H. W. Pearson* 8149

I. acanthoclada Dinter Fed. Rep. **18**, 428 (1922) could be this species. The type, *Dinter* 3048 from Kanus, 18°40′E, 27°50′S, just south of the Karasberg, has probably been destroyed. The very inadequate description could equally well apply to *I. circinnata*.

Subsect. C.3.f. **Brevipatentes** subsect. nov.

Type species *I. suaveolens* Jaub. & Spach.
Centre of distribution N.E. Trop. Africa.

Suffrutices ± argentei regiones aridas incoli. Folia parva, 1–5-foliolata. Inflorescentia brevis, 1–5-flora, axe quam legumen breviore. Legumina patentia, alia erecta, alia horizontalia, alia subreflexa, longitudine latitudine minime 4-plo excedente, minime 4-sperma, endocarpio (ubi viso) maculato.

Stipules up to c.6 mm. long, with a broad scarious base, attached to the petiole for
  c.2 mm., and, with it, persistent after the leaflets have fallen; leaflets 1–5, the
  lateral, where present, alternate . . . . 1. *ruspolii.*

Stipules not as above; lateral leaflets, where present, opposite:
 All leaves compound:
  Leaves digitately 3-foliolate; a small blister at each side of each leaflet and
   beneath its tip . . . . 2. *ogadensis.*
  Leaves 5-foliolate or pinnately 3-foliolate, without blisters:
   Hairs on stems closely appressed; petiole more than half as long as lateral
    leaflet . . . . 3. *suaveolens.*

Hairs on stems ± spreading at the tips, petiole less than half as long as lateral leaflet . . . . 4. *nebrowniana.*

Some at least of the leaves simple:

Leaves, or the larger ones, over 4 mm. wide; a short peduncle clearly discernible bearing 3–5-flowers (or traces of fallen flowers) . . . 5. *bainesii.*

Leaves not above 2 mm. wide:

Peduncle 1–2 mm. long bearing 1–2 flowers; stipules not or hardly discernible; outside of standard densely appressed-silvery-strigose in bud
6. *pseudointricata.*

Peduncle absent, flowers 1–2 in leaf axils on separate pedicels; stipules *c.*1·5 mm. long; outside of standard puberulent . . . . . . 7. *tanaensis.*

C.3.f.1. **I. ruspolii** *Bak.f.,* Vierteljahrschr. Nat. Ges. Zur. **49,** 192; L.T.A. No. 49.

Br. Somaliland; Ethiopia, Ogaden, Warandab *Keller* s.n. Z holotype, etc.; Somalia.

E. G. Baker's description of the leaves as digitately trifoliolate is wrong. They are 1–5-foliolate, the lateral leaflets alternate.

C.3.f.2. **I. ogadensis** *Gillett* sp. nov.

Species *I. ruspolii* aspectu similis, sed stipulis minutis, foliis digitatim trifoliolatis, foliolisque, ut in *I. brevicalyce* pustulatis, bene distincta.

Suffrutex ramis junioribus rectis, rigidis, argenteis, strigis appressis dense vestitis, ramis vetustioribus brunnescentibus. *Folia* digitatim 3-foliolata, aut ad ramos novos longiores distantia, aut ad ramos vetustiores glomerata; stipulae anguste deltoideae, argenteae, *c.*0·5 mm. longae; petiolus sparse strigosus 1–3 mm. longus; foliola, 0·25 mm. petiolulata, obovata basin angustata, utrinque sparse strigosa, strigis *c.*0·5 mm. longis, perlucide 3-pustulata, pustula una ad marginem quemquem, tertia subtus ad apicem. *Axis* racemi ad 6 mm. longus, pedunculo 1–2 mm. longo incluso, flores 3–5, legumina 1–2 ferens; pedicellus fructiferus 0·5–1·5 mm. longus, vel suberectus, vel horizontaliter patens vel subreflexus. *Calyx* strigosus, *c.*1·2 mm. longus, ad mediam divisus, lobis pustulatis. *Corolla* extus pubescens, *c.*4 mm. longa. *Filamenta* 3·5 mm. longa, antheris ellipticis, basin submuticis, ad 0·6 mm. longis, mucrone apicale fusco ad 0·2 mm. longo incluso. *Ovarium* pubescens, stylo glabro, basin complanato, 1·5 mm. longo, 1 mm. infra stigma punctiforme rectangulare curvato. *Legumen* cylindricum sparse strigosum, ad 17 mm. longum, 2 mm. latum, 7–9-spermum, endocarpio brunneo-maculato, maculis subconfluentibus. *Semina* compresse-rhomboidea, haud foveolata.

Ethiopia, Ogaden, S.E. of Uardere, 45°40′E, 6°45′N, *c.*600 m., rare, 1953, *Dr. P. Ellis* 60 holotype.

C.3.f.3. **I. suaveolens** *Jaub. & Spach,* Ill. Pl. Or. t. 489 (1856); F.T.A. **2,** 80; sensu Taubert in Engl. Pfl.welt O. Afr. C, 310 (1895); Bak.f., J. Bot. **41,** 240 (1903) et L.T.A. No. 110 pro minore parte; non sensu Brenan, Check List Tanganyika trees and shrubs 427, nec sensu Cronquist, F.C.B. **5,** 160.

I. sparsiflora Hochst. (nomen).

I. kerensis Chiov. in Pirotta Fl. Eritrea, Ann. Ist Bot. Rom. **8,** 417 (1908); L.T.A. No. 82.

Fr. Niger Col., Aouderas, Aïr, 8°50′E, 17°40′N, *Buchanan* s.n. BM; A.-E. Sudan, Darfour, *Purdy* 72, *Pfund* 16, 165, 267 *Lynes* s.n. BM; Eritrea, *Pappi* 6427 BM, *Pappi* 6253 (348), *Pappi* 6293 Fi holotype of *I. kerensis*; Ethiopia, North, *Schimper* (1854) 2288 *I. sparsiflora*, 1221 P holotype.

Pappi 6253K was distributed as *I. suaveolens* var. *subquadriflora* Chiov. but there has evidently been some mistake, as, from the characters which Chiovenda gives, the plant is clearly the true *suaveolens* and not his var. *subquadriflora* (*I. vohemarensis*). Although, when Schimper first collected this species and *I. vohemarensis*, Hochstetter distinguished them as *I. sparsiflora* and *I. subquadriflora* (without publishing either name), they have been considerably confused for the last sixty years. The difference may be summarised as follows:

I. suaveolens	*I. vohemarensis*
1. Peduncle 1–6 mm.	Peduncle 8–20 mm.
2. Fruits erect, spreading or reflexed.	Fruits erect, or nearly so.
3. Calyx lobes longer than tube, *c.*2 mm. long; calyx half as long as corolla.	Calyx lobes as long as or shorter than tube, *c.*1 mm. long; calyx much less than half as long as corolla.
4. Young twigs densely appressed silvery.	Indumentum on young twigs much sparser.
5. Leaflets 1–5.	Leaflets 3–9.

C.3.f.4. **I. nebrowniana** *Gillett* sp. nov.

"*I. variabilis N.E. Brown*" sensu N.E. Brown in Burtt Davy, Man. Flor. Transvaal **2**, 367 (1932) pro maj. parte; non J. Bot. **41**, 192 (1903); nec Kew Bull. **1909**, 103; nec L.T.A. No. 83.

I. bainseii aspectu similis; foliolis 3–5, nec 1–3; apicibus loborum calycis crasse glandulosis; pilis ramorum subpatentibus; pedunculo quam calycem saepius multo longiore; rhachide foliorum fere semper ultra foliola lateralia prolongata, differt. Ab *I. suaveolente* pilis ramorum subpatentibus, petiolis brevissimis, lobisque calycis differt.

Suffrutex semimetralis, ramis argenteis, pilis ± patentibus. *Folia* 3–5-foliolata; stipulae subulatae *c.*1 mm. longae; rhachis 6–9 mm. longa, petiolo 1–3 mm. longo incluso, saepius ultra foliola lateralia *c.*1 mm. prolongata, pilis subpatentibus subsparse strigosa; foliola suboblonga, basin attenuata apice rotundata, subrecurvata, pilis appressis *c.*0·4 mm. longis utrinque subsparse strigosa, ad 9 mm. longa, 3·5 mm. lata. *Axis* racemi [4]- 6–9–[13] mm. longus, pedunculo 2–4 mm. longo incluso, flores 3–5, fructos saepius 1–2 ferens; bracteae mox deciduae; pedicelli *c.*1 mm. longi, post anthesin vel erecti vel patentes vel subreflexi. *Calyx* strigosus, tubo 0·5 mm. longo, lobis crassis, subcylindricis, apice glandulosis, inaequalibus, superioribus *c.*0·4 mm., lateralibus *c.*0·5 mm., inferiore *c.*0·8 mm. longo. *Corolla* mox decidua, extus pubescens; vexillum ovatum [ad 4·5 mm. longum]; alae glabrae; carina calcarata, haud rostrata, [ad 4·5 mm. longa]. *Stamina* omnia fertilia, filamentis 3·5 mm. longis, antheris ovalibus, basin muticis, 0·5 mm. longis, mucrone apicale fusco 0·1 mm. longo incluso. *Ovarium* stylo curvato longius. [*Legumen* subcomplanatum, subcostatum, strigis *c.*0·5 mm. longis sparse strigosum, ad 1 mm. longum, 2 mm. latum, 1 mm. crassum, 4–7-spermum, endocarpio maculato, regione infra costam medianam excepta.]

Central Transvaal, Pietersburg distr., Pyramid estate 1365 m., in thornveldt bush clumps, flowers red, 26.2.1921, *Galpin* 8953 K (PRE). Waterberg distr. nr. Warmbath, 1110 m., Jan. 1906, *H. Bolus* 11797 K, BM, (BOL). Rustenberg distr., nr. Rustenberg, flowers red, 50 cm. tall, 6.2.1929, *Hutchinson* 2938. Pretoria distr., Magalisberg nr. Wonderboom, Dec. 1904, *Burtt Davy* 2667 K (PRE); on main road 50 km. N. of Pretoria, sweet bush veldt, 15.2.1946, *Codd* 822 K (PRE); Aapies river *Zeyher* 487; Hammanskraal, roadside, 29.12.1953, *A. D. J. Meeuse* 9122 K, EA, (PRE); Magalisberg in stony places, 6.11.1893, *Schlechter* 3646 K holo. (Z), BM isotypes.

As mature fruits are lacking in the type, they are described from *Hutchinson* 2938.

C.3.f.5. **I. bainesii** *Bak.*, F.T.A. **2**, 71 (1871); L.T.A. No. 19.

I. variabilis N.E. Brown in Bak.f., J. Bot. **41**, 192 (1903) et Kew Bull. **1909**, 103; L.T.A. No. 83; sensu N.E. Brown in Burtt Davy Man. Flor. Transvaal 2, 367 (1932) pro min. parte. Non *I. variabilis* De Wild. (1904).

"*I. simplicifolia Lam.*" sensu Bremekamp & Obermeyer, Ann. Trv. Mus. **16**, 419 (1935), non Lam.

TROP. BECHUANALAND, S. of Lake Ngami, *Baines* s.n. holotype. Kwebe hills, 1020 m., small bush *c.*70 cm. tall, *Lugard* 99, 119 syntypes of *I. variabilis* N.E. Br. Seshumo valley, *c.*25°15′E, 17°55′S, *Holub* s.n. Kaotwe 23°12′E, 22°30′S *Van Son.* in TRV 28915 BM; TRANSVAAL, Matebe valley N. of Zeerust, *c.*26°20′E, 25°30′S, *Holub* 736. *c.*5 km. N. of Ons Hoop, 870 m., *c.*27°42′E, 23°32′S, on sandy flats in low bushveldt, shrublet *c.*70 cm. tall, 23.2.1954, *Codd* 8447 K (PRE).

The areas occupied by *I. bainesii* and *I. nebrowniana* appear not to overlap, the former occupying a larger drier, rather lower area to the north-west of the comparatively restricted range of *I. nebrowniana*.

C.3.f.6. **I. pseudointricata** *Gillett* sp. nov.

"*I. intricata Boiss.*" sensu Balfour f., Bot. Socotra, Trans. R. Soc. Ed. **31**, 74 (1888); non Boiss., Fl. Or. **2**, 190 (1872); nec sensu Hutch. & Bruce, Kew Bull. **1941**, 118.

*Suffrutex c.*70 cm. altus, ramis rigidis argenteis, pilis appressis *c.*0·5 mm. longis densissime vestitis. *Folia* simplicia involuta plerumque deflexa, densissime pilis argenteis appressis vestita, 1·5–3 mm. longa *c.*0·5 mm. lata (in statu involuta: an post pluvios explicescentia?); stipulae minutissimae; petiolus 0·6 mm. longus. *Inflorescentiae* argenteae, 1–2–florae, pedunculo *c.*2 mm. longo, pedicello *c.*1 mm. longo, post anthesin suberecto—sub reflexo, bracteis *c.*0·2 mm. longis. *Calyx* argenteus ad 1·5 mm. longus fere ad mediam divisus, lobis deltoideis inaequalibus, inferiore quam alii paulo majore. *Vexillum* extus appresse strigosum, suborbiculare. *Alae* glabrae, suboblongae, unguiculatae. *Carina* calcarata, haud rostrata, extus pubescens. *Stamina* omnia fertilia, filamentis 4–5 mm. longis, antheris ellipticis, basin muticis, *c.*0·5 mm. longis, mucrone apicale purpureo 0·1 mm. longo incluso. *Ovarium c.*3 mm. longum, appresse strigosum, *c.*7-ovulatum, stylo basin curvato *c.*1·3 mm. longo, stigmate 0·1 mm. lato. *Legumen* subcylindricum strigosum, strigis appressis, *c.*0·4 mm. longis, ad 14 mm. longum, 1·5 mm. latum, *c.*5-spermum, endocarpio maculato. *Semina* flava, rotundato-oblonga, *c.*1·6 mm. longa.

SOCOTRA, Tamarid, 70 cm. tall, "fl. carneo", vern. name SIDERI, 13.4.1881, *Schweinfurth* 298 holotype. Very abundant on the plains, especially Habidu plain, Feb.-March 1880, *Bayley Balfour* 107 pro majore parte. Hadibo plain, 15–45 m., dominant at this alt., everywhere in the coastal plain, native name MUSAHAT, 22.1.1953, *G. Popov* SO/28 BM, EA.

On the Kew sheet of *Balfour* 107 there are 2 sterile branches of a plant with much longer (up to 11 mm.) erect leaves and longer hairs (*c.*0·7–0·8 mm.) which probably belong to some other species.

The resemblance between this species and *I. intricata* Boiss No. C.3.p.2, which has pinnate leaves, alternate leaflets, much longer, more-flowered inflorescences and deflexed fruits is superficial and due to convergence.

C.3.f.7. **I. tanaensis** *Gillett* sp. nov.

Species *I. pseudo-intricatae* similis; ut in clavi distinguitur.

Suffrutex 20 cm. altus, ramis junioribus rigidis, argenteis, pilis appressis brevibus densissime vestitis, vetustioribus atro-brunneis. *Folia* simplicia, complicata, cuneata, apice paulo recurvata, appresse dense strigosa, saepe ad nodos conglomerata, ad 5 mm. longa, petiolo 0·5 mm. longo incluso, 1·5 mm. lata; stipulae anguste deltoideae ad 1·5 mm. longae. *Flores* in axillis foliorum solitarii vel binati, pedicellis *c*.0·7 mm. longis; pedunculus nullus. *Calyx* argenteus, tubo 1·2 mm. longo, lobis subulatis, basin haud expansis, *c*.0·6 mm. longis. *Vexillum* extus pubescens ellipticum, apice emarginatum, *c*.5 mm. longum, 3·5 mm. latum. *Alae* glabrae ± dolabriformes, breviter unguiculatae, *c*.3·2 mm. longae. *Carina* ad marginem pubescens, calcare 1 mm. longo, haud rostrata, 4·5 mm. longa. *Stamina* omnia fertilia, filamentis 4·5 mm. longis, antheris ellipticis *c*.0·6 mm. longis, mucrone apicale brunneo incluso. *Ovarium* glabrum, 3·5 mm. longum, 12-ovulatum, stylo glabro, e base attenuato, in medio curvato, *c*.2 mm. longo, stigmate globoso *c*.0·1 mm. diametro. *Legumen* haud visum.

Kenya, near the Tana river, "Burra 10 miles inland from Koro Koro", *c*.39°52′E, 0°32′S, 105 m., on red sandy soil only. In full flower after recent showers, 8.4.1934, *H. C. Sampson* 66 holotype.

Subsect. C.3.g. **Pilosae** subsect. nov.

Type species, *I. pilosa* Poir.

Centre of distribution, South Tropical Africa.

Pili saepius longi, patentes, albi vel fulvi (haud brunnei vel atri). Folia simplicia vel pauce- (ad 5-) foliolata, saepius illis §§**Dissitiflorarum** majora. Stipulae saepius longiores latioresque quam aliter in genere (**Latestipulatis** exceptis). Racemus axillaris pauce- *c*.20-florus. Calyx profunde fissus, lobis mediam staminum excedentibus. Legumina recta, erecta vel paulo patentia. Endocarpium maculatum. Semina inter se haud distantia.

I. glabra L. from India apparently belongs to this subsection.

Leaves simple:
 Petiole 4–10 mm. long . . . . 1. *petiolata*.
 Petiole 0–3 mm. long:
  Leaves ± oblong with long spreading hairs:
   Stems rather stout (up to 3 mm.); plant ± erect; calyx lobes not gland tipped; peduncles short (1–5 mm.); pods short (up to 5 mm.) and broad (*c*.2·5 mm.), *c*.2-seeded; endocarp with small ± confluent spots
2. *kirkii*.
   Stems more slender (up to 1·5 or 2 mm. diam.); plant spreading or procumbent; calyx lobes often gland-tipped; peduncles longer (*c*.8–13 mm.); pods 10–15 mm. long, *c*.1·5 mm. wide, *c*.6-seeded; endocarp with 2 clear lines of distinct spots on each valve . . . . 3. *erythrogramma*.
  Leaves broadly obovate with short, thick, appressed hairs . . . 4. *nummularia*.
Leaves 3–5-foliolate, occasionally some of them simple:
 Leaf rhachis, including petiole, over 5 mm. long:
  Petiole of 3-foliolate leaves longer than the rest of the rhachis, a pair of stalked glands on its lower half; leaflets suborbicular, but little longer than wide
5. *biglandulosa*.
  Petiole of 3-foliolate leaves usually shorter than rest of rhachis, without stalked glands; leaflets oblong, usually twice as long as wide:
   Inflorescence 1–4-flowered; hairs whitish; leaflets often ± acute; pods 10–12-seeded . . . . 6. *pilosa*.
    Pods pubescent . . . . α. var. *pilosa*.
    Pods glabrous when mature . . . . β. var. *angolensis*.
   Inflorescence 5–20-flowered; hairs yellowish; leaflets ± obtuse; pods shorter, pilose, 7-9-seeded . . . . 7. *fulvopilosa*.

Leaf rhachis, including petiole, under 2 mm. long, leaves subdigitately trifoliolate, almost sessile, inflorescence 2–4-flowered . . . . 8. *torrei*.

C.3.g.1. **I. petiolata** *Cronq.*, Bull. J. Bot. Et Brux. **22**, 221 (1952); F.C.B. **5**, 132.

B. Congo, *Van Meel in De Witte* 5303 BR holotype.

C.3.g.2. **I. kirkii** *Oliv.* in Hook. Ic. Plant. t. 1416 (1883); L.T.A. No. 29.

I. preladoi Harms, Engl. B. Jahrb. **26**, 284 (1899); L.T.A. No. 28.

I. smithioides R.Vig., Not. Syst. **13**, 368 (1949).

Tanganyika, T6 *Kirk* s.n. holotype, T8; Port. E. Afr. (Nr. Beira, Apr. 1895, *Schlechter* s.n. B†?; Cabeceira Grande *Prelado* 20 B†? syntypes of *I. preladoi*), Mozambique, 5.4.1894 *O. Kuntze* s.n.; Madagascar, west coast, *Perrier* 1435 P, 1435 bis P, 4875 P syntypes of *I. smithioides*, etc.

O. Kuntze s.n. was identified by Harms as *I. preladoi* in a footnote to his original description and may be taken as its neotype if both syntypes prove to have been destroyed. *A. R. Torre* 1256 from Nampula, Port. E. Afr. has pointed leaves up to 4·5 cm. long and calyx lobes *c*.4 mm. long. It may represent a variety of this species.

C.3.g.3. **I. erythrogramma** *Welw. ex Bak.*, F.T.A. **2**, 73 (1871); L.T.A. No. 30; F.C.B. **5**, 132. Including var. *pauciflora* Bak.f. & Martin ex De Wild., Bull. J. Bot. Et Brux. **8**, 151 (1923).

B. Congo, Bequaert 7194 BR holotype of var. *pauciflora*; Tanganyika, T8 *Schlieben* 6114 BM; Port. E. Afr., *Schlechter* 12173 K, BM; Nyasaland; S. Rhodesia; Angola, *Welwitsch* 2005, 2007 syntypes.

Rather variable; the calyx lobes are usually gland-tipped in the western part of its range but less often so in the east. In this respect it resembles *I. pilosa*.

C.3.g.4. **I. nummularia** *Welw. ex Bak.*, F.T.A. **2**, 74 (1871); L.T.A. No. 27.

Angola, *Welwitsch* 2009 K, BM type.

C.3.g.5. **I. biglandulosa** *Gillett* sp. nov.

I. pilosae affinis; ut in clavi distinguitur.

Herba, ut videtur annua, ramis procumbentibus, *c*.40 cm. longis, pilis patentibus ad 1·5 mm. longis vestitis. *Folia* pinnatim trifoliolata, vel rarius 1- vel 5-foliolata, ubique pilosa; stipulae anguste lanceolatae, apice glandulosae, 3–4 mm. longae; petiolus 10–15 mm. longus, infra mediam glandulis duobus *c*.0·7 mm. stipitatis praeditus; rhachis ultra foliola 5–6 mm. prolongata; stipellae filiformes, *c*.0·6 mm. longae, mox caducae; petioluli 1·5 mm. longi; foliolum terminale quam lateralia duplo majus, elliptico-orbiculare, *c*.20 mm. longum, 16 mm. latum. *Inflorescentiae normales* racemi pilosi, *c*.10–12 flori, *c*.10–12 mm. longi, pedunculo *c*.6 mm. longo incluso; bracteae lineares apice glandulosae ad 2 mm. longae; pedicelli *c*.0·8 mm. longi. *Inflorescentiae reductae* in axilla folii infimi quique rami, saepe unifoliolati vel ad bracteam glandulosam reducti, orientes, vel 2–4-florae breviter pedunculatae vel uniflorae vel floribus 1–2 abortivis. *Calyx* pilosus, tubo 0·5 mm. longo, lobis aequalibus, anguste triangularibus, apice glandulosis, 2 mm. longis. *Vexillum* extus pubescens, obovatum, basin late cuneatum, 2·8 mm. longum, 2·2 mm. latum. *Alae* glabrae, dolabriformes, breviter unguiculatae, 3 mm. longae, 1·2 mm. latae. *Carina* ad margines et ad commissuram pilosa, 0·4 mm. calcarata, haud rostrata, 2·8 mm. longa. *Stamina* omnia fertilia, filamentis 1·5–2 mm. longis, antheris basin muticis, late oblongis, 0·4 mm. longis, appendice apicale fusco, 0·06 mm. longo incluso. *Ovarium* ± dense pilosum, *c*.2 mm. longum, parte superiore nonnunquam sterile, *c*.4-ovulatum (an nonnunquam pluries?), stylo glabro, basin curvato, *c*.0·9 mm. longo, stigmate oblique capitato *c*.0·15 mm. diametro. [*Legumen* pubescens *c*.9–13 mm. longum, 1·5 mm. latum, 4–7-spermum] Semina haud visa.

A.-E. Sudan, Umdurain S. Kordofan, Corolla red, 28.12.1907, *H. M. & A. F. Broun* 1353; Tanganyika, T4, Ujiji, 660 m., on sand near lake, "vine", flower pale pink outside, deep pink inside, March 1939, *Mrs. M. V. Loveridge* 705 holotype.

A remarkable species: *Broun* 1353 is a poor specimen with the reduced type of inflorescence alone in evidence; it does however show fruits. *Loveridge* 705, while flowering copiously does not show even immature fruits.

C.3.g.6. **I. pilosa** *Poir.* in Lam. Encyc. Suppl. **3**, 151 (1813); L.T.A. No. 90 p.p. excl. var. *multiflora*; Brenan in Mem. N. York Bot. Gardn. **8**, 251 (1953).

α. var. **pilosa.**

I. guineensis Schum. & Thonn., Beskr. Guin. Pl. 367 (1829).

Hemispadon pilosus Endl., Flora **15**, 385 (1832), t.1.

Senegal (*Kohaut* s.n. type of *Hemispadon pilosus*); Port. Guinea; Fr. Sudan, *Hagerup* 359 BM; Gold Coast (? *Thonning* s.n. P. holotype), *Thonning* s.n. C. holotype of *I. guineensis*; Nigeria; Br. Cameroons; A.-E. Sudan; Eritrea.

β. var. **angolensis** *Bak.f.*, J. Bot. **41**, 243 (1903).

Angola, *Welwitsch* 2022 holotype, 4149, *Gossweiler* 779.

C.3.g.7. **I. fulvopilosa** *Brenan*, Mem. N. York Bot. Gdn. **8**, 250 (1953); F.C.B. **5**, 140.

I. pilosa Poir. var. *multiflora* Bak.f., J. Bot. **41**, 243 (1903); L.T.A. No. 90 var.

Sierra Leone; Gold Coast; Nigeria; B. Congo; A.-E. Sudan, *Wyld* 473 BM; Uganda, U1–2, U4; Tanganyika, T4, T7–8; Port. E. Afr., Vila Cabral, *Mendonç*a 688 K, LISC, Guruè Mts. *Torre* 1521 LISC, Zobuè *Torre* 5802 K, LISC; Nyasaland, *Buchanan* 292, 413, 824 syntypes; S. Rhodesia, Salisbury, *Davies* 489, 700; N. Rhodesia, Abercorn, *Mrs. Richards* 5086, 5449, etc., Ndola *Fanshawe* 1036.

C.3.g.9. **I. torrei** *Gillett* sp. nov.

Species propter indumentum, foliaque fere sessilia subdigitatim trifoliolata, inflorescentiamque paucifloram, fructosque erectos, ab congeneribus omnibus bene distincta.

Suffrutex ramosus *c.*60 cm. altus, ramis purpureis dense patente pilosis. *Folia* omnino, sed superne sparsius, pilosa, fere sessilia, subdigitatim trifoliolata, vel rarius superiora 2-vel 1-foliolata; stipulae subulatae, rubrae, pilosae, *c.*3 mm. longae; petiolus ad 0·5 mm. longus; petioluli 0·5 mm. longi; rhachis ultra foliola lateralia 0–0·5 mm. prolongata; foliola cuneato-obovata, apice apiculata, 6 mm. longa, 3 mm. lata, costa superne impressa, nervis lateralibus utrinque 2–3 subtus bene distinctis. *Racemi* axillares 2–4-flori, ad 2 cm. longi, pedunculo pubescente 3–10 mm. longo incluso; bracteae 0·7 mm. longae mox deciduae; pedicelli pubescentes *c.*1 mm. longi, post anthesin erecti. *Calyx* pilosus, tubo 0·5 mm. longo, dentibus 3–3·5 mm. longis. *Vexillum* extus pubescens, suborbiculare, apiculatum, *c.*5 mm. longum, 4·5 mm. latum. *Alae* glabrae, 3·3 mm. longae, unguiculo 0·2 mm. longo incluso. *Carina* extus pubescens, haud rostrata, laterale 1 mm. calcarata, in toto 5 mm. longa. *Stamina* omnia fertilia, filamentis 4–5 mm. longis, antheris subfuscis, ellipticis, basin muticis, 0·7 mm. longis, mucrone 0·2 mm. longo incluso. *Ovarium* *c.*6-ovulatum, 2 mm. longum, in stylum glabrum 3 mm. longum, paulo supra mediam abrupte rectangulare curvatum, sensim attenuatum; stigma minutum globosum. *Legumen* 3–5-spermum, ± erectum, pubescens, subtetragonum, 9–13 mm. longum, 2·2 mm. latum, 2·2 mm. crassum, apice acutum 0·5 mm. rostratum, endocarpio maculato. *Semina* matura haud visa.

Port. E. Afr., Limpopo distr., 150–180 km. from Camicado on Sauté road, 22°40′S, 33°20′E, savannah on white sandy soils, 15.5.1948, *A. R. Torre* 7837 K iso LISC holotype.

Subsect. C.3.h. **Viscosae** *Rydberg*, N. Am. Flor. 24, 140 (1923).

Types species *I. colutea* (Burm.f.) Merr. (*I. viscosa* Lam.).

Central of distribution, Central Africa.

Leaves pinnate, lateral leaflets almost always opposite. Multicellular hairs, usually gland-tipped, present. Biramous hairs white. Racemes usually lax, the fruits usually spreading (neither erect nor deflexed) but if deflexed having glandular hairs. Calyx often more than half as long as stamens. Endocarp usually spotted.

A difficult group containing several ill-defined polymorphic species.

Fruits $\pm$ erect:
 Perennials:
  Leaflets 1–4-jugate, strictly opposite; inflorescence rarely over 5 cm. long; stamens *c*.5 mm. long . . . . 1. *hilaris.*
  Leaflets up to 8-jugate, often subalternate; inflorescence often over 10 cm. long; stamens 7–10 mm. long . . . . 2. *sordida.*
 Annuals; inflorescence 1–3 cm. long:
  Filaments *c*.4 mm. long; petiole $\pm$ half as long as the basal leaflets; multicellular hairs copious, up to 0·5 mm. long 10δ. *colutea* var. *dembianensis.*
  Filaments under 3 mm. long; petiole as long as, or longer than, the basal leaflets; multicellular hairs few and not over 0·2 mm. long . . . .
3. *milne-redheadii.*

Fruits $\pm$ spreading, or sometimes subreflexed:
 Rhachis often not prolonged beyond last leaflet pair: inflorescence up to 5 times as long as subtending leaf; calyx less than half as long as the 5 mm. long stamens:
  Leaflets 9–15; glandular hairs usually copious . . . . 4. *heterotricha.*
  Leaflets 5–7; glandular hairs often scanty . . . . 5. *lydenbergensis.*
 Rhachis usually definitely prolonged beyond last leaflet pair: inflorescence rarely twice as long as subtending leaf:
  Fruit sharply curved in a semicricle, covered with glandular hairs 6. *curvata.*
  Fruit $\pm$ straight:
   Inflorescence much more than 5 mm. long; fruits usually over 3 mm. long:
    Fruit *c*.1–1·7 wide:
     Corolla rarely as much as 3 times as long as calyx:
      Petiole usually shorter than basal leaflets:
       Leaflets less than 3 times as long as wide, rarely more than 4-jugate, finely appressed strigose above; glandular hairs and epidermis of stem usually $\pm$ reddish-purple; fruit not over 12 mm. long the endocarp not, or obscurely, spotted . . . 7. *mimosoides.*
       Pod usually more than 2-seeded; multicellular hairs not restricted to calyx:
        Glandular hairs of uneven length (the longer often over 1 mm.) often found sparsely on the leaf rhachis, but not on the leaflets, except sometimes, on the midrib . . . .
α. var. *mimosoides.*

        Glandular hairs very numerous, all short (not over 0·3 mm.) often on the leaflets, especially their margins . . . .
β. var. *viscidior.*

        Pod 1–2-seeded; multicellular hairs often restricted to calyx
γ. var. *brachycarpa.*

Leaflets at least 3 times as long as wide, often 5-jugate; glandular
hairs short (not over 0·3 mm.); fruit up to 15 mm. long the
endocarp strongly spotted . . . . 10 α. *colutea* var. *dembianensis*.

Petiole usually longer than the basal leaflets (if shorter, leaflets
over 11) leaflets often 11 or more; endocarp spotted:

Fruits having glandular hairs:

Glandular hairs very numerous, short (not over 0·3 mm.),
extending on to the leaflets, usually coloured, like the
epidermis of the stems, ± reddish-purple:

Leaflets 3–11, rarely over 8 mm. long, glabrous or with
glandular hairs only above, fruit almost without biramous
hairs . . . . 8. *adenoides*.

Leaflets 9–17, often over 8 mm. long, appressed-strigose
above; fruit closely covered with biramous, as well as
glandular hairs . . . . 9. *viscidissima*.

Glandular hairs, if all short, sparse, or absent from leaflets, if
dense and present also on leaflets then some of those on
stems long (over 1 mm.) like the stem epidermis usually
yellowish green, inflorescence not markedly secund . . . .
10. *colutea*.

Filaments not above 4 mm. long:

Glands on stiff, spreading, multicellular hairs:

Leaflets with glandular hairs on their margins, rarely
more than 3 times as long as wide; glandular hairs
on stems often long . . . . α. var. *colutea*.

Leaflets without glandular hairs, usually more than
3 times as long as wide, filaments c. 2 mm. long
β. var. *linearis*.

Glands subsessile; leaflets *c.* twice as long as wide . . . . . .
γ. var. *somalensis*.

Filaments 5–7 mm. long . . . . ε. var. *grandiflora*.

Fruits and leaflets free from glandular hairs, which are often very
sparse elsewhere:

Inflorescence markedly secund; leaflets more than 11 . . . . . .
11. *barteri*.

Inflorescence not secund; leaflets not more than 11:

Leaflets much more than twice as long as wide; medifixed
hairs appressed; glandular hairs, where present, usually
long . . . . 12. *dyeri*.

Leaflets hardly more than twice as long as wide; medifixed
hairs ± spreading; glandular hairs, where present short
13. *delagoaensis*.

Corolla *c.*4 times as long as calyx; leaflets 3-7, linear; petiole 10–20 mm.
long, stipellae well developed, *c.*1 mm. long; glandular hairs
numerous, short, purple tipped, absent from leaflets   14. *hewittii*.

Fruit *c.*2 mm. wide:

Pod greenish or brown, over 8 mm. long; stems prostrate, somewhat
zigzag, densely covered with glandular hairs; leaflets glabrous or
sparsely strigose above . . . . 15. *bangweolensis*.

Glandular indumentum on stems shorter than their diameter;
endocarp not spotted . . . . α. var. *bangweolensis*.

Glandular indumentum on stems longer than their diameter; endocarp spotted . . . .   β. var. *lacusvictoriae*.

Pod silvery, stem ± ascending:

Inflorescence dense, secund, many-flowered; pods 4–6 mm. long
with black glandular hairs as well as a dense silvery coat of
medifixed hairs . . . . 16. *secundiflora*.

Stems without long red spreading multicellular hairs . . . .
α. var. *secundiflora*.

Stems ± covered with long red spreading multicellular hairs . . .
β. var. *rubripilosa*.

Inflorescence lax, not secund, with 3–12-flowers only, pod up to
15 mm. long, the glandular hairs yellowish, very short and sparse
17. *argentea*.

Inflorescence compact, not above 5 mm. long; fruits 2-seeded, *c.*3 mm. long
18. *conferta*.

C.3.h.1. **I. hilaris** Eckl. & Zeyh., Enum. Pl. Afr. Austr. 241 (1836); L.T.A.
No. 134; F.C.B. **5**, 133.

I. microscypha Bak., Kew Bull. **1897,** 255; L.T.A. No. 131.

I. nyikensis Bak., Kew Bull. **1897,** 254; L.T.A. No. 128.

I. patula Bak., Kew Bull. **1897,** 255; L.T.A. No. 100.

I. wentzeliana Harms, E.B.J. **30,** 326 (1901); L.T.A. No. 133.

I. moeroensis De Wild., Ann. Mus. Congo Ser. **5**: **1**, 133 (1904); L.T.A. No. 107.

I. zigzag De Wild., Fed. Rep. **11**, 538 (1913); L.T.A. No. 129.

I. hockii De Wild & Bak.f., Fed. Rep. **2**, 297 (1913); L.T.A. No. 130.

"*I. heterocarpa Welw. ex Bak.*" sensu R. E. Fries, Wiss ergeb. Schwed. Rhod.
Kongo. Exped. 79 (1914) non Welw. ex Bak.

I. similis N. E. Brown, Kew Bull. **1921,** 292.

I. hybrida N. E. Brown, Kew Bull. **1925,** 152.

S. Belg. Congo, *Hock* s.n. BR holotype of *I. hockii, Bequaert* 106 BR holotype of
I. zigzag, Rogers 26210 holotype of *I. similis, Verdick* s.n. BR holotype of *I. moeroen-
sis*; Tanganyika, T4, T7 (*Goetze* 1268B† holotype of *I. wentzeliana*); Port. E.
Afr., Niassa Prov. *Pedro & Pedrogao* 3546 EA, Delagoa bay; Nyasaland, *Whyte*
s.n. holotype of *I. patula, Whyte* 151 holotype of *I. nyikensis Whyte* s.n. holotype of
I. microscypha; S. Rhodesia; N. Rhodesia, *R. E. Fries* 455, 1319, etc.; Transvaal,
Mrs .Pott 5079 holotype of *I. hybrida*; Orange Free State; Natal; E. Cape Prov.,
Ecklon & Zeyher 1605 isotype.

This species is a perennial with a woody rootstock from which leafy shoots spring
up following fires. Such species are commonly very variable in Southern Africa and
I. hilaris is no exception. It varies especially in the width of the leaflets, the density
of the indumentum, the presence or absence of glandular hairs, in the length of the
inflorescence and in that of the calyx. There seems to be a centre of variation in
Nyasaland where *nyikensis, patula* and *microscypha* are glandular forms. It would
appear that the species may assume a very different aspect if it escapes grass fires
for a season or two and grows up into a bushy plant. *N. E. Brown* suggested that in
the Transvaal *I. hybrida* might be a result of hybridisation between *I. hilaris* and
some glandular species, and such an explanation might well account for the diversity
of *I. hilaris* s. lat. in Nyasaland also. Of 111 sheets of this species preserved at Kew
2 only show fruits.

C.3.h.2. **I. sordida** *Benth. ex Harv.*, Fl. Cap. **2,** 190 (1862).

S. RHODESIA, Charter district, *Harvey* in SRGH, 3634 SRGH; E. TROP. BECHUANALAND, *Holub* s.n.; TRANSVAAL, *Zeyher* 480, *Burke* 316 syntypes, etc.

Pole Evans in PRE 13102 from Messina, N. Transvaal appears to be this species but is completely without the usual glandular hairs.

C.3.h.3. **I. milne-redheadii** *Gillett* sp. nov.

Herba tenuis annua, saepius erecta, ad 25 cm. alta. Caules foliorumque rhaches pilis biramis albis appressis *c.*0·5 mm. longis haud dense vestiti; pili multicellulares glandulosi brevissimi perpauci etiam adsunt. *Folia* 3–17-foliolata; stipulae lineares *c.*1 mm. longae; rhachis, petiolo ad 8 mm. longo incluso, ad 25 mm. longa, ultra foliola lateralia *c.*1 mm. prolongata; foliola oblongo-lanceolata, basin attenuata, apice ± rotundata, apiculata, subtus pilis validis ad 1 mm. longis strigosa, superne pilis tenuioribus ad 0·2 mm. longis strigosula, ad 10 mm. longa et 2·5 mm. lata. *Racemi* axillares, 3–10-flori, pedunculo strigoso 4–7 mm. longo incluso, ad 3 cm. longi; bracteae lineares, mox caducae, *c.*0·7 mm. longae; pedicelli *c.*1 mm. longi, post anthesin erecti vel paulo patentes. *Calyx* strigosus, tubo 0·7 mm. longo, lobis subulatis aequalibus, apice atrato-rubris, *c.*1·7 mm. longis. *Vexillum* ovatum, 2·7 mm. longum, 2·6 mm. latum, extra albo-strigosum. *Alae* glabrae, auriculatae, 2·5 mm. longae, 1 mm. latae. *Carina* ad suturam pilosa, laterale 0·5 mm. calcarata, erostrata, 2·5 mm. longa. *Stamina* omnia fertilia, filamentis 2–2·3 mm. longis, antheris atro-rubris, 0·3 mm. longis, apice mucronatis, basin muticis. *Ovarium* strigosum 1·8 mm. longum, stylo glabro fere recto 0·7 mm. longo, stigmate globoso 0·2 mm. diametro. *Legumen* ascendens vel patens, rectum, dense et breviter strigosum, 6–9-spermum, 10–15 mm. longum, *c.*2 mm. latum, 1·3 mm. crassum, endocarpio haud intense maculato. *Semina* paulo foveolata, aspectu laterale rectangulata, aspectu terminale lenticularia, 0·8 mm. longa, 1·1 mm. lata. 0·6 mm. crassa.

TANGANYIKA, Ufipa district; Ilemba-Kasamvu, 1200–1500 m., common on the rocky slope of the scarp, flowers pink, 21.3.1950, *Bullock* 2691. Songea district; Songea airfield, 1020 m., by path through secondary *Brachystegia-Uapaca* woodland, on red soil probably overlying "laterite". Calyx pale green, the tips of the lobes reddish, standard and wings pale cerise, an irregular whitish area flecked with red at base of standard, keel mainly greenish white, 14.3.1956, *Milne-Redhead and Taylor* 9155; Unangwa hill, *c.*6 Km. E. of Songea, 1140 m., shallow soil by rocks on steep slope in *Brachystegia* woodland, standard salmon with a white-marked brick red zone at base, salmon-pink outside, wings cerise, keel greenish white, cerise at upper edge, 22.3.1956, *Milne-Redhead and Taylor* 9323, Songea-Mbamba bay road, near upper crossing of Luhekea river, 630 m., on steep shady bank in Brachystegia-Uapaca woodland, 4.4.1956, *Milne-Redhead and Taylor* 9518; Lower slopes of Litenga hill *c.*40 Km. W. of Songea, 100 m., *Brachystergia-Uapaca* woodland, unripe fruits grey green, 19.4.1956, *Milne-Redhead and Taylor* 9766.

NYASALAND, Malosa, Zomba, in *Brachystegia manga* woodland, 10.3.1955, *G. Jackson* 1496. N. RHODESIA, Western Province; Solwezi district; quarry just N. of Mutanda bridge, dry stony waste ground, 21.6.1930, *Milne-Redhead* 564 type; Shibenda's village, weed in native garden by pool, 26.7.1930, *Milne-Redhead* 785. Ndola, wayside, 7.3.1954, *Fanshaw* 946.

This species is related to *I. subargentea* De Wild. *I. microcephala* Bak.f. *I. colutea* (Burm.f.) Merrill and *I. dyeri* Britten. It differs from the first in its smaller flowers, longer petioles, broader leaflets and in consistently having a few short glandular hairs. From *I. microcephala* it differs in the smaller flowers, fewer flowered, shorter, inflorescences, suberect fruits and narrower leaflets. The same characters and the scarcity and shortness of its glandular hairs separate it from *I. colutea* except that the var. *linearis* of that species has equally small flowers. *I. dyeri* has fewer leaflets, rather larger flowers, longer inflorescences with spreading fruits and fewer seeds, and its multicellular hairs, when present, are long.

C.3.h.4. **I. heterotricha** *DC.*, Prod. **2**, 227 (1825); L.T.A. No. 137.

I. rudis N.E. Br., Kew Bull. **1925**, 149.

I. heterotricha var. *rhodesiana* Bak.f., J. Bot. **41**, 262 (1903).

S. RHODESIA, *Rand* 62 BM holotype of var. *rhodesiana*, etc.; S.W. AFRICA; N. CENTRAL CAPE & BRIT. BECHUANALAND, *Burchell* 2635 istotype; TRANSVAAL, *Rogers* 19485 holotype of *I. rudis*.

This species varies considerably but the different forms pass into one another so gradually that it is hardly feasible to distinguish them.

C.3.h.5. **I. lydenbergensis** *N. E. Brown*, Kew Bull. **1925**, 149.

PORT. E. AFR., Lourenco Marques between Moamba and Ressano Garcia, *Torre* 2205 K, LISC; TRANSVAAL, *Wilms* 310 holotype, etc.

Torre 2205 like *Pole Evans* 1567 from the Pretoria district represents a form of this species which is free from the usual glandular multicellular hairs.

Besides *I. sordida*, *I. heterotricha*, *I. lydenbergensis*, *I. adenoides* and *I. colutea* a number of other species usually possessing glandular hairs which would come into §§**Viscosae** as here defined occur in the tropical and subtropical parts of the Union of South Africa and Angola. Among these are *I. adenocarpa* E. Mey, *I. pechuelii* O. Kuntze, *I. pearsonii* Bak.f. *I. dolichothyrsa* Bak.f., and *I. enormis* N.E. Br. These species are difficult to define and little understood. *Torre* 7322 LISC from between Caniçado and Nalazi, Limpopo distr. Port. E. Afr. falls in this group but does not match anything at Kew. Until the whole group has been more thoroughly investigated than is now possible it seems inadvisable to describe it as new.

C.3.h.6. **I. curvata** *Gillett* sp. nov.

I. coluteae affinis, legumine curvato, setis glandulosis obsito, distinguitur.

Herba annua, *c.*25 cm. alta, ± erecta, ramis pilis medifixis appressis *c.*1 mm. longis et pilis glanduligeris erectis ad 0·4 mm. longis, subsparse vestitis. *Folia* 9–13-foliolata; stipulae subulatae, ad 4 mm. longae, sparse pilosae; rhachis pilis medifixis appressis et pilis glandulosis brevissimis obsita, ultra foliola lateralia ad 3 mm. prolongata, in tota ad 35 mm. longa, petiolo 9–12 mm. longo incluso; stipellae subnullae; petioluli 0·5 mm. longi, foliola anguste obovata vel elliptico oblonga, terminale aliis haud majore, apice rotundata, eglandulosa, pilis medifixis appressis, inferne *c.*1 mm. longis, superne brevioribus tenuioribus, ± sparse vestita, ad 12 mm. longa et 4 mm. lata. *Racemus* 10–20-florus, axe pilis glandulosis brevibus et pilis medifixis obsito, ad 25 mm. longo, pedunculo 5–7 mm. longo incluso; pedicelli *c.*0·7 mm. longi, post anthesin vel subreflexi vel suberecti; bracteae haud visae, mox deciduae. *Calyx* et pilis medifixis et pilis glandulosis obsitus, tubo 0·7 mm. longo, lobis aequalibus acuminatis apice subulatis *c.*1·8 mm. longis. *Vexillum* extus sparse pubescens, ellipticum, 3·2 mm. longum, 1·6 mm. latum. *Alae* glabrae, 0·4 mm. unguiculatae, lamina suboblonga, 2 mm. longa, 0·5 mm. lata. *Carina* extus sparse pubescens, 0·5 mm. calcarata, haud rostrata, 3·4 mm. longa. *Filamenta* *c.*2·5 mm. longa; antherae haud visae. *Ovarium* 1·8 mm. longum, pilis medifixis et pilis glandulosis ad 0·7 mm. longis dense obsitum *c.*13-ovulatum, stylo incurvato, crasso, *c.*1 mm. longo. *Legumen* valde arcuatum, subtorulosum, pilis medifixis et pilis glandulosis ± dense obsitum, *c.*17 mm. longum, 1·4 mm. latum, 0·9 mm. crassum, 9–12-spermum, endocarpio valde maculato.

TANGANYIKA, T7, Rungwe distr. Mbozi Ipata, 9°2′S, 32°58′E, 1200 m., fairly common, 8.4.1932, *R. M. Davies* 560 EA holotype K isotype.

C.3.h.7. **I. mimosoides** *Bak.*, F.T.A. **2**, 90 (1871); L.T.A. No. 94.

α. var. **mimosoides**.

I. affinis De Wild., Bull. J. Bot. Et Brux. **8**, 144 (1923); L.T.A. No. 204: non Harvey.

I. brevipetiolata Cronquist, F.C.B. **5,** 145 (1954).

I. shirensis Taub. ex Bak.f., L.T.A. 116, No. 78 (1926).

"*I. viscosa Lam.*" sensu Brenan, Mem. N.Y. Bot. Gdn. **8,** 251; non Lam.

BR. CAMEROONS, Bamenda, *Johnstone* 182 FHI; FR. CAMEROONS, *Mildbraed* 10231; B. CONGO, East, *Bequaert* 7014 BR holotype of *I. affinis* De Wild and *I. brevipetiolata*; ETHIOPIA, *Schimper* (1844) 1475, (1863–8) 1540, *Gillett* 14499; UGANDA, U2, U4; KENYA, K3, K5; TANGANYIKA, T2, T7; PORT. E. AFR., *Sousa* 1281, *Pedro & Pedrogao* 3663 EA; NYASALAND, *Buchanan* 152 BM, 694 BM, 1491 BM. These were labelled *I. shirensis* by E. G. Baker and one of these should be considered the neotype of that name if the Berlin sheet of *Buchanan* s.n. (sketch at BM) has been destroyed.

S. RHODESIA, *Teague* 26, 49, 50, *Eyles* 7095, 8931, *Gilliland* 224 SRGH, *Chase* 903 SRGH, *Wild* 44521; N. RHODESIA, *Mrs. Richards* 859, 4967, 5552; ANGOLA, *Welwitsch* 2045 BM isosyntype, 4151 BM, *Gossweiler* 9485.

β. var. **viscidior** *Gillett* var. nov.

Pili glandulosi ad ramos, rhachides foliorum, inflorescentiasque breves (ad 0·3 mm. longi) numerosissimi. Foliola saepe etiam pilis glandulosis plus minusve vestita. Endocarpium minute et obscure brunneo-maculatum.

TANGANYIKA, T7 Mbeya distr. Usafwa 1650 m., herbaceous, common in scrub, 24.3.1932, *R. M. Davies* 149, K holo., EA isotype ibid. 65 cm. tall, *Davies* 134 K, EA. Ukinga and Ubena area Njombe distr. March 1937, *E. J. Ward* U18 K, EA. Lower plateau N. of Lake Nyassa *J. Thomson* s.n.

γ. var. **brachycarpa** *Gillett* var. nov.

A varietatibus duobus aliis *I. mimosoidei* ovario 2-ovulato fructuque 1–2-spermo differt. Pili multicellularii purpurei satis sparsi et saepe ad calycem solum reperti. Filamenta *c.*3 mm. longa calycem vix superantia.

TANGANYIKA, Songea district: Lipumba, 1200 m., on hard gravelly red clay loam of abandoned road, 28.4.1933 *Stenhouse* 5 K, holo, EA isotype. 6·5 Km. S. of Peramiho, 1050 m., *Brachystegia-Uapaca* woodland on red loam, perennial from a woody base, standard and wings bright cherry red, keel and outside of standard grey tinged pink, apex of keel deep pink, 3.4.1956, *Milne-Redhead and Taylor* 9511. *c.* 5 Km. E. of Songea, 1080 m., *Brachystegia-Uapaca* woodland on red loam by roadside, short lived perennial, 20.3.1956, *Milne-Redhead and Taylor* 9306. Matengo hills, Mbinga, 1400 m., bushland, 24.2.1936, *Zerny* 460. Matengo hills, Mpapa, 1470 m., bushland, annual or short lived perennial, 5.3.1956, *Milne-Redhead and Taylor* 8979. Miyau, 1620 m., 6.3.1956, *Milne-Redhead and Taylor* 8979/A.

I. mimosoides varies considerably and it is with some hesitation that all the plants cited above are included under var. *mimosoides.* The type of *I. brevipetiolata* is more robust than usual; it closely resembles *Purseglove* 595, 614, 3534 from S.W. Uganda. *I. mimosoides* var. *rhodesica* Bak.f. L.T.A. 880 (1930) differs considerably from the rest of the species. The type *Eyles* 4484 BM, SRGH, which is without fruit, has alone been seen. Until further material is available the status of this plant is uncertain. *I. grata* E. Mey from Natal is closely related: it has narrower leaflets and fruits densely covered with glandular hairs, which are absent from the rest of the plant. *Milne-Redhead* 4448 from N. Rhodesia which has very numerous, very long, multicellular hairs on the stems and longer petioles may represent a distinct variety or even species. It does not show fruits.

C.3.h.8. **I. adenoides** *Bak.f.* Vierteljahrschr. Nat. Ges. Zur. **49,** 182 (1904); L.T.A. No. 116.

S. RHODESIA, Matopos, *Flanagan* 3136 BM; TRANSVAAL, *Rehmann* 6217 (Z holo.) K, BM isotypes, etc.

C.3.h.9. **I. viscidissima** *Bak.*, F.T.A. **2**, 100 (1871); L.T.A. No. 234.

TANGANYIKA, T7 *Davies* 549 K, EA; NYASALAND, *Nicholson* s.n., *Jackson* 1571; S. RHODESIA, *Wild* 1810, *Jack* 155 SRGH; N. RHODESIA, *Mrs. Richards* 4467, *Milne-Redhead* 4503; ANGOLA, *Welwitsch* 2042 isotype, etc.

Davies 111 from Mbosi distr. S.W. Tanganyika has the long petiole of this species and is intermediate between it and *I. mimosoides.*

C.3.h.10. **I. colutea** (*Burm.f.*) *Merrill*, Phil. J. Sci. **19,** 355 (1921).

α. var. **colutea.**

Colutea siliquosa enneaphyllos Indiae orientalis, siliquis & foliis aversa parte pubescentibus Pluk., Phytographia t. 166 f. 3 (1691).

Galega colutea Burm. f., Fl. Ind. 172 (1768); non sensu Willd. Sp. Pl. **3,** 1246 (1803).

Tephrosia colutea (Burm.f.) Pers., Syn. **2,** 329 (1807) quoad nomen solum.

Indigofera viscosa Lam., Encyc. **3,** 247 (1789); L.T.A. No. 103; F.C.B. **5,** 144.

I. glutinosa Perr. ex DC., Prod. **2,** 227 (1825) non Vahl, nec Schum.

I. consanguinea Klotzsch, Pet. Moss. Bot. 50 (1862).

I. seticulosa Harv., Fl. Cap. **2,** 196 (1862).

I. zenkeri Harms ex Bak.f., J. Bot **41,** 241 (1903); L.T.A. No. 117.

I. zenkeri var. *brevifoliolata* De Wild., Bull. J. Bot. Et Brux. **8,** 158 (1923).

I. multifoliolata De Wild., Pl. Beq. **2,** 539 (1924).

I. junodii N.E. Br., Kew Bull. **1925,** 147.

I. viscosa Lam. var. *brachycarpa* Bak.f., L.T.A. 124 (1926).

YEMEN; PAKISTAN; INDIA, *unknown collector*, BM figured in Pluk. t. 166 f.3; CEYLON; INDONESIA; AUSTRALIA, *Armstrong* 385 holotype of *I. seticulosa*, wrongly thought to be from S. Africa; CAPE VERDE ISLES; SENEGAL, *Perrottet* 182 (P holo.) BM isotype of *I. glutinosa* Perr. ex DC; FR. SUDAN, *Hagerup* 243 BM; NIGERIA; FR. CAMEROONS, *Zenker* 1456 holotype of *I. zenkeri*; B. CONGO. *Bequaert* 3371 BR holotype of *I. multifoliolata, Bequaert* 6097 BR holotype of *I. zenkeri* var. *brevifoliolata*; A.-E. SUDAN; ERITREA; ETHIOPIA; SOMALIA, *Senni* 243 FI; SOCOTRA; UGANDA, U2–4; KENYA, K1, K3–7; TANGANYIKA, T1–7; ZANZIBAR; PORT. E. AFR. (*Peters* s.n. B † type of *I. consanguinea*); NYASALAND; S. RHODESIA; N. RHODESIA; ANGOLA, *Gossweiler* 400 type of *I. viscosa* var. *brachycarpa*; S.W. AFRICA; TRANSVAAL, *Junod* 645 holotype of *I. junodii.*

β. var. **linearis** *Bak.f. ex Gillett* var. nov.

A var. *colutea* foliolibus saepius paucioribus (ad 11), angustioribus, eglandulosis, pilis glandulosisque ramorum brevioribus differt. Filamenta 2–2·5 mm. longa.

ETHIOPIA, Gafta, 16.9.1838, *Schimper* 1205; KENYA, K1, *Dandu*, 750 m., *Gillett* 13164; K4, Kibewezi, 870 m., *Dummer* 4581; K7 Kiunga Is. *O.U. Exped.* L.I.; TANGANYIKA, T2, Moshi, 900 m., *Haarer* 28B, 780 m., *Haarer* 576, 592; T3, Mazinde, 435 m., *Drummond & Hemsley* 2340; T5, Mpapwa, *Mrs. Hornby* 121; NYASALAND, *Jackson* 1089; N. RHODESIA, Abercorn distr., *Mrs. Richards* 1192, 4506, 4507, 4571, 5040; Mwinilunga distr. *Milne-Redhead* 4403; S. RHODESIA, Salisbury distr. 1140 m. *Eyles* 4483 BM holo. SRGH; isotype, *Davies* 491 SRGH, *Aylen* in SRGH 11839 K; Nyamandhlovu distr. 1170 m., *Sharp* in SRGH 19737 K; Hartley distr., 1200 m., sand veldt, *R. M. Hornby* 2866 K, SRGH; Marandellas distr. *Dehn* 632 K, SRGH; Victoria distr., 1200 m., sandveldt, *Robinson* 306 K, SRGH.

Extreme forms of this variety could be confused with *I. hewitii* which differs in having the calyx much shorter in relation to the corolla and the stipellae much better developed.

γ. var. **somalensis** (*Bak.f.*) *Cufod.*, Bull J. Bot. Et Brux. **25**, 263 (1955).

I. viscosa var. *somalensis* Bak.f. J. Bot. **41**, 242 (1903); L.T.A. p. 124.

ETHIOPIA, Ogaden, Milmil, *Donaldson Smith* s.n. BM holotype; BRIT. SOMALI-LAND, Balleh Murca Bugh, a few km. N. of Omr Aji, *c.*8°25′N, 46°17E, 720 m., 25.10.1954, *Bally* 10196. *Glover & Gilliland* 290 from Wardere wells, Ogaden is intermediate between vars. α and γ.

δ. var. **dembianensis** (*Chiov.*) *Cufod.* 1.c.

I. viscosa var. *dembianensis* Chiov., Ann. Di Bot. **9**, 58 (1911); L.T.A. p. 124.

ETHIOPIA, Dembia (the region N. of Lake Tana), chiefly around Gondar, *Chiovenda* 1229 FI, 1328 FI, 1670 FI, 2278 FI, 2391 FI. *Schimper* (1840) 1475.

A well marked variety represented, on the 5 sheets listed, by 13 entire plants. They differ from the rest of the species as follows:

	var. α	var. β	var. γ	var. δ
1.	Gland hairs short and long ..	short	very short indeed	short
2.	Gland hairs present on leaflet midrib, lamina and margin	absent from leaflets or on midrib only	as in α	as in β.
3.	Petiole longer than basal leaflets	as in α	as in α	Petiole about half as long as basal leaflets.
4.	Axis of infructescence often exceeding 3·5 cm.	rarely as much as 3·5 cm.	as in α	not above 2·8 cm.
5.	Flowering and fruiting pedicels spreading	as in α	as in α	flowering and fruiting pedicels stiffly erect.
6.	Stamens 2–4 mm. long ..	2–3 mm.	c.3 mm.	4 mm.
7.	Fruits up to 20 mm. long ..	up to 15 mm.	?	12–15 mm.

It is probable that var. *dembianensis* will eventually be considered a distinct species, but the study of further material from the area is advisable before a decision is reached on this point. There is an affinity between var. *dembianensis* and *I. subargentea* De Wild.

ε. var. **grandiflora** *Gillett* var. nov.

A var. *colutea* floribus majoribus differt. Calyx 2–3 mm. longus. Stamina 5–7 mm. longa, haud 3–4 mm. ut in var. *colutea.*

KENYA, Furroli, 3°42′N, 38°0′E, *c.*1000 m., lava plain, semi desert—*Commiphora-Acacia* scrub transition, 14.9.1952, *Gillett* 13855. Lake Elmenteita, 1800 m., rocky lake shore with soda present, abundant, 17,8.1948, *Bogdan* 1869 holotype, ibid. Soysambu estate, rocky slopes, frequent, 30.7.1947, *Bogdan* 937. Rift valley 64 Km. S.W. of Nairobi on Magadi Rd., 1050 m., rocky slopes, frequent, 20.7.1947, *Bogdan* 864. Lorgasailie, 1140 m., in semi desert grassland after rain. 20.8.1947, *H. Andrews in Bally* 5180. Magadi, 660 m., on dry alkaline soil between rocks, June 1947, *J. G. Williams in Bally* 5231.

TANGANYIKA, Masai district, Oldonya Lengai *c.*35°55′E, 2°45′S, 1350–2100 m., *St. Clair Thompson* 198, 289. The Ngaserai just off Arusha—Namanga Rd., *c.*2° 53′S, 1200 m., dry grass plain on coarse granitic sands, 19.4.1955, *P. Huxley* 202. Arusha district *c.*3 Km. S. of Ol Doinyo Sambu, 1840 m., dry stony ground by roadside, near maize field, prostrate, leaves grey-green, tinged purplish, flowers conspicuous, the wings equal and opposed to the standard, bright cerise, 1.11.1955, *Milne-Redhead and Taylor* 7161.

The typification of *Galega colutea* has caused some difficulty. Prof. C. Baehni has very kindly searched the Geneva herbarium where Burmann's collections are preserved, but could find no specimen that might be the type. N. L. Burmann in his flora Indica was attempting to account for, and give binomials to, all plants up till then described from India. As Merrill says in his account of this work (Phil J. Sci. **19**: **3**, 329 (1921)), it is clear that, in some cases, Burmann had to work from published descriptions and figures alone, without seeing specimens. Unfortunately he does not state when he is, and when he is not, using specimens. The descriptive phrase which Burmann adds to that of Plukenet runs as follows "leguminibus racemosis strictis adscendtibus rugosis, foliolis obovatis sericeis". There is nothing in this which could not be derived from Plukenet's figure. The words "rugosis" and "sericeis" accord ill with our present species, but may readily be understood as misinterpretations of Plukenet's figure by someone who had seen no specimen. There is thus every reason to believe that Burmann's name is based wholly on Plukenet's figure, which must thus be its type. The figure in turn is based on a specimen in the Sloane herbarium in the British Museum which is undoubtedly conspecific with *Indigofere viscosa* Lam. Botanists therefore, under the present rules, have no option but to accept Merrill's combination in place of Lamarck's long established name. Willdenow, as is clear from his description misapplied Burmann's name to a totally different plant, which cannot possibly be reconciled with Plukenet's figure. Persoon included Willdenow's plant in his new genus *Tephrosia*.

C.3.h.11. **I. barteri** *Hutch. & Dalz.*, F.W.T.A. 1, 391 (1928); Kew Bull. **1929,** 16; L.T.A. 880.

Gold Coast, *Morton* in GC 6197, 25017; Nigeria, *Barter* 951 holotype, 1604, *Lely* P645, *Dent Young* 47.

Andrews, Fl. Pl. A.-E. Sudan **2,** 217 records this species from Equatoria. In the absence of fruits it is not clear that the specimen concerned, *Wyld* 246 BM may not be a form of *I. secundiflora*.

C.3.h.12. **I. dyeri** *Britten*, J. Bot. **35,** 453 (1897); L.T.A. No. 101.

I. macra Bak., Kew Bull. **1897,** 255; non E. Mey (1836).

I. nematopoda Bak.f., Viertjahrschr. Nat. Ges. Zur. **70,** 216 (1925) fig. 3; L.T.A. No. 102.

Kenya, K7 *Mainwaring* 3040 (somewhat approaching *I. microcephala*); Port. E. Afr., *Faulkner* 74 and 420, *Kirk* s.n., etc.; Nyasaland, *Whyte* s.n. holotype, *McClounie* s.n., etc.; S. Rhodesia, *Fisher* 1404 S.R.G.H. *Holub* s.n. BM fragment, Z holotype of *I. nematopoda, Saunders Davies* s.n. BM; N. Rhodesia, *Sandwith* 73, *Jack* 198, 198/A, *Wild* 1821; E. Transvaal, Lions Creek, 300 m., *Schlechter* 12190.

As here understood this species is polymorphic. Some specimens, *e.g. Faulkner* 74 and 420 have copious long glandular hairs on the stem. In most specimens such hairs are very few, while *Holub* s.n. is totally without them. Most plants seem to be slender annuals but *Jack* 198, 198/A are stouter and may be perennials.

C.3.h.13. **I. delagoaensis** *Bak.f. ex Gillett* sp. nov.

Herba perennis, ramis patentibus, pilis medifixis crispis dense et pilis multi-cellulosis glandulosis brevibus erectis ± sparse vestitis. *Folia* 5–11-foliolata; stipulae subulatae, ad 4 mm. longae; rhachis ultra foliola lateralia 2–4 mm. pro-longata, indumento eo ramorum simile, ad 35 mm. longa, petiolo 8–10 mm. longo incluso; stipellae haud visae; petioluli *c.*1·4 mm. longi; foliola aequalia, oblongo-elliptica, apiculata, utrinque pilis medifixis crispis vestita, ad margines etiam pilis glandulosis paucis obsita, *c.*8 mm. longa, 4 mm. lata. *Racemi* 3–15 cm. longi, multiflori, pedunculo quam partem floriferam multo breviore, indumento eo ramorum simile; bracteae brunneae, pilosae, anguste lanceolatae, *c.*2 mm. longae, deciduae; pedicelli 0·5–1 mm. longi, post anthesin patentes vel subreflexi. *Calyx* pilis biramosis multis et pilis glandulosis paucis vestitus, tubo *c.*1 mm. longo, lobis subulatis apice tenuissimis inaequalibus, superioribus 1·7 mm., lateralibus 1·9 mm., inferiore 2·1 mm. longo. *Vexillum* extus pubescens, ± oblongum, *c*.3·2 mm. longum, 2·6 mm. latum. *Alae* glabrae, 0·1 mm. unguiculatae, suboblongae, 2·3 mm. longae, 1 mm. latae. *Carina* extus pubescens, 0·5 mm. calcarata, haud rostrata, 3·3 mm. longa. *Stamina* omnia fertilia, filamentis 2·5–3 mm. longis, antheris late ellipticis basin muticis, *c*.0·5 mm. longis, appendice apicale fusco 0·1 mm. longo incluso. *Ovarium* strigosum 2 mm. longum, *c*.9-ovulatum; stylus curvatus, e base attenuato, *c*.1.5 mm. longus. *Legumen* subcylindricum, crispe strigosum, 13 mm. longum, 1·5 mm. latum, *c*.8-spermum, endocarpio maculato. *Semina* brunnea, subcuboidea, ± foveolata.

Port. E. Afr., Lourenco Marques district, *c*.60 m., Aug. 1928, *Rogers* 21372 BM; 30 m., 30.11.1897, *Schlechter* 11532 K holo. BM isotype; prostrate plant along hills, 25.10.1919, *H. L. Shantz* 334; Near Bela Vista, in sandy soil, perennial, 20.11.1940, *A. R. Torre* 2099 K, LISC. Near Chobela, Maguda, annual herb in *Acacia* forest, 30.12.1947, *A. R. Torre* 7015 LISC. Manhica, abundant, Aug. 1930, *Gomes Sousa* 450; in sandy places 30 m., Aug. 1896, *H. Bolus* 1139 K, BM, UPS; Polana, flowers pink, very frequent, 11.1.1920, *J. Boule* s.n.; Polana flats, shrub, 27.8.1940, *A. J. Hornby* 2060.

S. Rhodesia, Mtao distr. bright red flowers, Jan. 1923, *J. M. Rattray* 161 SRGH, Umvuma, Jan. 1923, *Eyles* 6969 K, SRGH.

Transvaal, Kruger nat. park, nr. Pretorius Kop 600 m., open grassy places in subtropical woodland, semi prostrate perennial, red flowers, 4.2.1949, *Codd & De Winter* 4923 K (PRE).

The S. Rhodesian specimens have unusually short leaves and a rather stiffer indumentum with no glandular hairs. They may represent a distinct taxon.

C.3.h.14. **I. hewittii** *Bak.f.*, J. Bot. **71**, 341 (1933).

I. dehniae Merxmueller, Trans. Rhod. Sci. Ass. **43**, 21 (1951).

N. Rhodesia, Mkushi distr., 13°55′S, 29°26′E, *Hewitt* 2 BM holotype.

S. Rhodesia, Marandellas distr., *Dehn* 168 (M holo.) SRGH isotype of *I. dehniae*, *Wild* 3297 K, SRGH; Rusapa distr. *Dehn* 168/ A/52 SRGH; Miami distr. *Wild* 1809; Victoria *Monro* 1275 BM; Matobo distr. *Miller* 2299.

C.3.h.15. **I. bangweolensis** *R. E. Fries*, Schwed. Rhod.-Kongo Exp. 78, t. 9, f.3. (1914); L.T.A. No. 138.

α. var. **bangweolensis.**

N. Rhodesia, Lake Bangweolo, *R. E. Fries* 798 UPS, 885 UPS syntypes; Lake Mweru *Bullock* 3820; Abercorn *Mrs. Richards* 1175, 4412, 5208, *Hutchinson & Gillett* 3850; Samfya *Fanshawe* 319.

β. var. **lacus-victoriae** *Gillett* var. nov.

A var. *bangweolense* pilis glandulosis ramorum longissimis (ad 5 mm.), endocarpioque maculato differt.

Tanganyika, T1 Ukerewe Island, decumbent, small red flowers, *N. V. Rounce* 24 K holo. EA isotype, ibid. *Conrads* 5143 EA, *Conrads* in EA 10369 K, EA.

This species varies greatly, especially var. *bangweolensis*. In the type the leaf rhachis is not, or hardly prolonged beyond the lateral leaflets, whereas in other specimens it is prolonged. The amount of glandular indumentum varies. *Fanshawe* 319 has an unusually short petiole (*c*.1–2 mm.) whereas in *Bullock* 3820 it is up to 13 mm.

C.3.h.16. **I. secundiflora** *Poir.* in Lam. Encyc. Suppl. **3**, 148 (1813); L.T.A. 202; F.C.B. **5**, 145.

α. var. **secundiflora.**

I. oligosperma, DC., Prod. **2**, 228 (1825).

I. glutinosa Vahl ex Schum., Beskr. Guin. Pl. 370 (1829) non Perr. ex DC.

SENEGAL (*unknown collector* s.n. P. holotype of *I. oligosperma*), etc.; FR. SUDAN, Koulikoro; *Chevalier* 3232 P.; GOLD COAST (*Thonning?* s.n. P. holotype) (*Thonning* s.n. C. type of *I. glutinosa* Vahl ex Schum.); FR. NIGER Col., *Hagerup* 513 BM; NIGERIA; A.-E. SUDAN; ETHIOPIA.

β. var. **rubripilosa** *De Wild.*, Bull. J. Bot. Et Brux. **8**, 156 (1923); F.C.B. **5**, 146.

I. bequaertii De Wild., Fed. Rep. **11**, 538 (1913).

I. secundiflora var. *gondarensis* Pichi Sermolli, Acc. Nazz. Lincei **7** missione al lago Tana **1**, 64, t. 11 (1951).

OUBANGUI, *Chevalier* 5703; B. CONGO, *Bequaert* 474 BR holotype of *I. bequaertii*, *Bequaert* 3365 BR holotype; A.-E. SUDAN; ERITREA; ETHIOPIA, *Pichi-Sermolli* 2265 isotype of var. *gondarensis*; UGANDA, U1–4; KENYA, K5; TANGANYIKA, T1, T4, T7; NYASALAND; N. RHODESIA.

In the type of var. *gondarensis* the long red multicellular hairs characteristic of var. β are even longer and denser than usual. There seems however no clear line between this and the rest of the variety. *Tisserant* 1290 P, the type of *I. secundiflora* var. *oubanguiensis* Tiss., Bull. Mus. Hist. Nat. Par. 2nd Ser. 3, 166 (1931) has more leaflets than usual and is more slender, approaching *I. barteri* except for the fruits, which are those of *I. secundiflora*. *Dekindt* 3043 P a syntype of *I. secundiflora* var. *schimperi* Tiss. ibid. p. 165 has no fruits. It does not seem clear that it belongs to this species.

C.3.h.17. **I. argentea** *Burm. f.*, Fl. Ind. 171. (1768); non L. (1771); L.T.A. No. 198.

I. semitrijuga Forsk. var. *tetrasperma* DC., Prod. **2**, 230 (1825).

"*I. semitrijuga Forsk.*" sensu Bak., F.T.A. **2**, 93, non Forsk.

I. burmannii Boiss, Fl. Or. **2**, 189 (1872).

?I. semitrijuga Forsk. var. *?macrocarpa* Vatke in Oest. Bot. Zeitschr. 1879: 221 (1879).

I. arenaria A. Rich. var. *strigosa* Terrac., Ann. ɪst Bot. Rom. **5**, 109 (1893).

EGYPT; HEJAZ; ADEN; OMAN; S. PERSIA (*Garcin* s.n. G ? holotype of *I. argentea*, *I. burmannii* and var. *tetrasperma*); PUNJAB; SIND; MAURITANIA, 21°N, *Monod* 10628 IFAN; FR. SUDAN; NIGER-CHAD, fide Tisserant; CHAD DISTR., Tibesti; A.-E. SUDAN; ERITREA (*Terraciano* s.n. Fɪ. type of *I. arenaria* var. *strigosa* fide Chiovenda in Pirotta Fl. Eritrea Ann. ɪst Bot. Rom. 8, 88 (1903)); BR. SOMALILAND; ETHIOPIA. Ogaden; KENYA, Teita district, N'dara (*Hildebrandt* 2399B type of var. *macrocarpa*).

The glandular hairs in this species are very short, and readily overlooked. The record for Kenya is doubtful as no specimen has been seen.

C.3.h.18. **I. conferta,** *Gillett,* Kew Bull. 1955, 579 (1956).

I. grisea Bak., F.T.A. **2**, 80 (1871) pro maj. parte, non Desv. (1826).
NIGERIA.

Subsect. C.3.i. **Centrae** subsect. nov.

Type species *I. ugandensis.*

Centre of distribution South Central Africa.

Pili multicellulares glandulosi desunt; pili biramosi albi vel rarius fulvi. Folia unifoliolata vel pinnata, rhachi ultra foliola lateralia prolongata. Inflorescentia laxa. legumina recta 3-vel ultra-sperma haud glabra, patentia. Calyx brevis vel longus. Endocarpium saepius maculatum.

A rather unsatisfactory group showing affinities in several directions.

Leaves compound:

Keel not rostrate, not longer than standard, stipules usually 2 mm. or more long:

Pod not above 1·5 mm. wide, sometimes over 15 mm. long, not torulose, the suture *c*.0·2 mm. wide; leaflets ± appressed-strigose above:

Indumentum on calyx and young inflorescence white:

Perennials woody at least at the base; leaves subtending inflorescences 1–5-foliolate; petiole 2–6 mm. long, usually longer than stipules:

Calyx less than one-third as long as filaments; leaflets rarely over 6 mm. long . . . . *lydenbergensis.*

Calyx one-third or more of length of filaments; leaflets often over 6 mm. long:

Leaves subtending inflorescences usually 1–2-jugate . . . . . *hilaris.*

Leaves subtending inflorescences 3–jugate; indumentum on fruit sparse . . . . 10. *heterocarpa.*

Annual, or, if perennial, leaflets more numerous, or petiole shorter than above:

Petiole usually less than 3 mm. long and shorter than stipules:

Leaflets not more than 3 times as long as wide; stems more or less reddish grey; erect half woody annuals up to *c*.1 m. tall:

Pods over 10 mm. long:

Calyx lobes about as long as tube, much shorter than filaments which are *c*.3 mm. long; bracts shorter than pedicel, often persistent; leaves 2–4-jugate . . . . 1. *quarrei.*

Calyx lobes longer than tube, rather more than half as long as the *c*.4 mm. long filaments; bracts longer than pedicels, caducous; infructescence often much more than 5 cm. long . . .
2. *ugandensis.*

Pods not above 10 mm. long; calyx lobes longer than tube, about half as long as filaments . . . . .3. *subulifera.*

Pods 1–2-rarely 3-seeded, filaments *c*.4 mm. long . . . .
α. var. *subulifera.*

Pods 4–6-seeded, filaments *c*.6 mm. long . . . β. var. *polysperma.*

Leaflets linear-lanceolate, more than 3 times as long as wide; calyx lobes much longer than tube, more than half as long as filaments; fruits 6– or more seeded, 15–20 mm. long; annual herb, the stems yellowish green . . . . 4. *subargentea.*

Petiole usually over 3 mm. long and longer than stipules:

Leaflets usually at least twice as long as wide; hairs on fruit dense, hiding the surface; infructescence rarely over 5 cm. long; annuals 20–40 cm. tall, the stems yellowish green:

Leaflets 9–19, pods usually over 10 mm. long. . . .5. *microcephala.*
Leaflets 3–9, pods rarely over 10 mm. long . . . . *dyeri.*

Leaflets less than twice as long as wide; stems often ± reddish; infructescence usually over 5 cm. long; indumentum on fruit less dense:

Calyx half as long as filaments, or less, the teeth about as long as the tube; hairs ± appressed; erect, *c*.1 m. tall . . . 6. *wildiana.*

Calyx as long as filaments, teeth much longer than tube; hairs ± crisped . . . . *delagoensis.*

Indumentum on calyx and young inflorescence golden . . . .
7. *phyllanthioides.*

Pod 2 mm. or more wide, rarely over 15 mm. long, often rather torulose, the
suture *c*.0·5 mm. wide; endocarp faintly spotted; leaflets often glabrous
above, or nearly so, or with lax spreading hairs . . . . 8. *vicioides.*

Indumentum ± appressed . . . . α. var. *vicioides.*

Indumentum spreading:

Pod spreading, not above 12 mm. long . . . . β. var. *rogersii.*

Pod ± ascending, *c*.14 mm. long . . . . γ. var. *occidentalis.*

Keel rostrate, longer than standard; stipules often under 2 mm. long; pods
10–12 mm. long; leaflets 5–9 . . . . .9. *livingstoniana.*

Leaves unifoliolate; perennials with stiff, spreading, often fulvous hairs:

Inflorescence shorter than subtending leaf; leaves up to 7 cm. long, ± narrowed
at the base, petiole up to 7 mm. long; branches erect or spreading . . . . . .
11. *hermannioides.*

Inflorescence longer than subtending leaf; leaves up to 4 cm. long, usually
truncate or cordate at the base; petiole up to 4 mm. long; branches prostrate
12. *taylori.*

C.3.i.1. I. quarrei *Cronquist*, Bull. J. Bot. Et Brux. **22**, 222 (1952); F.C.B. **5**, 165.

B. CONGO, South (*Quarré* 1067 BR holotype) *Hirschberg* 241, *Rogers* 10979;
N. RHODESIA, Solwezi distr., *Milne-Redhead* 752, Mwinilunga distr., *Milne-
Redhead* 3863, Kitwe *Fanshawe* 715, 2270.

C.3.i.2. I. ugandensis *Bak.f.*, L.T.A. 155, No. 211 (1926).

UGANDA, U4 *Dümmer* 694 BM, 3192 holotype, 4501 *Chandler* 1549; TANGANYIKA,
T4 *Rounce* 42 K, EA.

C.3.i.3. I. subulifera *Welw. ex Bak.*, F.T.A. **2**, 96 (1871); L.T.A. No. 209;
F.C.B. **5**, 156, non sensu Tisserant, Bull. Mus. Hist. Nat. Par. 2nd ser. **3** (1931)
167, 271.

α. var. **subulifera.**

I. mounyinensis Tisserant, l.c. 168 pro parte.

B. CONGO, *Quarré* 3162, *Ritschard* 1803; NYASALAND, *Jackson* 1143, 1573;
S. RHODESIA, Marandellas distr., *Corby* 650 SRGH; N. RHODESIA, Fort Jameson
distr., *Exell, Mendonça & Wild* 1138, Abercorn distr. *Mrs. Richards* 5231; ANGOLA,
Welwitsch 2049 K lectotype BM, P isolectotypes 2048 b K, P, pp. BM pp *Gossweiler*
2520.

β. var. **polysperma** *Gillett* var. nov.

I. mounyinensis Tisserant pro. parte.

A var. subulifera ovulis seminibusque pluribus (4–6, haud *c*.2) et floribus
majoribus (filamentis *c*.6 mm. haud *c*.4 mm. longis) differt.

ANGOLA, Huilla between Catamba and Hay March 1850, *Welwitsch* 2050 K,
holotype BM, P isotypes; Mounyino *Antunes & Dekindt* (3179 P) 3102 P.

Baker described the pods of *I. subulifera* as being usually 2-, casually 1-seeded.
The pods which he saw are those of *Welwitsch* 2048b which is a mixed number
including, in the Paris and British Museum sheets some *I. mildbraediana* q.v as
well as the species here considered. When Tisserant investigated this species he had
besides the Welwitsch material, *Antunes & Dekindt* 3102 and 3179 of which the
former shows only flowers and the latter (which I have not seen) presumably the
4–5-seeded fruits which he describes. Tisserant came to the conclusion that the

71

several-seeded fruits are normal and the 2-seeded fruits, which Baker describes, an abnormality due to abortion. However, a dissection of Welwitsch's flowering material shows that, while No. 2050 has 4–5 ovules and might thus be expected to produce fruits such as Tisserant describes, No. 2049 has only 2 ovules and should thus produce fruits such as those of 2048 without abortion. Furthermore there is a difference in flower size, the filaments being 6 mm. long in No. 2050 and only 4 mm. long in 2049, 2048 and all the other specimens cited as var. *subulifera*. These facts seem best summed up by the recognition of two varieties and are certainly not in accordance with Tisserant's abortion hypothesis.

I. subulifera varies considerably in leaflet size and shape. It is closely related to *I. mimosoides* from which it differs in the absence of glandular multicellular hairs and in the shorter rather broader fruits (1·6 instead of 1·2 mm. wide).

C.3.i.4. **I. subargentea** *De Wild.*, Bull. J. Bot. Et Brux. **8**, 157 (1923); L.T.A. No. 115; F.C.B. **5**, 168.

I. kandoensis Bak.f., Rev. Zoo. Bot. Afr. **21**, 301 (1932).

B. Congo (*Bequaert* 4915 BR holotype), (*De Witte* 118 BR holotype of *I. kandoensis*); Uganda, U2, U4; Kenya, K3, K5; Tanganyika, T4, T7; N. Rhodesia, Abercorn distr.

C.3.i.5. **I. microcephala** *Bak.f.*, L.T.A. 116, No. 77 (1926).

I. shinyangensis Milne-Redhead, Kew Bull. **1936**, 471; F.C.B. **5**, 168 pro majore parte, plantis pilis brunneis exclusis.

B. Congo, East, *Scaetta* 1379; Uganda, U4 *Maitland* 591 holotype, *Chandler-Hancock* 138, 1709; Tanganyika, T1 *Bax* 152 holotype of *I. shinyangensis*, *Koritschoner* 2103 EA, *Bax* 148, 150, *Burtt* 3779 BM, 5147, *R. L. Davis* 258, *Welch* 166, *Staples* 437 K, EA; T7 *R. M. Davies* 83, 89, 314; N. Rhodesia, *Kirk* s.n., *Mrs. Richards* 4527, 4635, 4776.

The epithet *microcephala* is misleading, as the "small heads" into which the budding inflorescences are condensed lengthen into very lax racemes as the fruiting stage approaches. The species varies considerably. The type of *I.'shinyangensis* is a stout plant with unusually broad leaflets. At first sight it appears distinct, but intermediates between it and the type of *I. microcephala*, which is an unusually small plant, are more numerous than either extreme. The species is closely related to *I. dyeri*, differing in the large fruits and more numerous leaflets. *Mainwaring* 3040 from near Mombasa is intermediate. It is also very close to *I. subargentea* and some of the plants cited above from T7 and N. Rhodesia are ± intermediate. From Uganda and T1, on the other hand, intermediates have not been seen.

C.3.i.6. **I. wildiana** *Gillett* stat. et nom. nov.

I. grata E. Mey var. *longeracemosa* Bak.f., L.T.A. 151 No. 195 (1926), non *I. longeracemosa* Boiv. ex Baill.

Suffrutex ad 1·3 m. altus, ramis subrubris, sparse appresse strigosis. *Folia* [5]-7-9-[13]- foliolata; stipulae subulatae *c.*1·5 mm. longae; rhachis appresse strigosa, ultra foliola lateralia *c.*3–5 mm. prolongata, ad 40 [50] mm. longa, petiolo 5–7 [–12] mm. longo incluso; foliola oblongo-elliptica, utrinque appresse strigosa, apice rotundata vel subacuta, basin subattenuata, *c.*9 mm. longa, 4 mm. lata. *Racemi* multiflori, post anthesin ad 14 cm. elongati, pedunculo 5 mm. longo incluso; bracteae strigosae, lanceolatae, erectae, mox deciduae, ad 1·5 mm. longae; pedicelli *c.*1 mm. longi, post anthesin patentes. *Calyx* strigosus, tubo obliquo superne *c.*1 mm. longo, inferne *c.*1·8 mm. longo, lobis subulatis, superioribus 1·2 mm. longis, inferiore *c.*1·4 mm. longo, sinibus, ut in *I. emarginella* inaequalibus, illo inter lobos superiores quam 4 alios duplo latiore. *Vexillum* ovatum extus strigosum, apice apiculatum, *c.*5·5 mm. longum, 4·5 mm. latum. *Alae* glabrae, dolabriformes, *c*.0·2 mm. unguiculatae, *c*.4·5 mm. longae. *Carina* extus strigosa, *c*.1 mm. calcarata, haud rostrata, *c*.5·5 mm. longa. *Stamina* omnia fertilia, filamentis 4·5–5 mm. longis, antheris ovatis, basin muticis, 0·7–0·8 mm. longis, appendice apicale fusco 0·2–0·3 mm. longo incluso. *Ovarium* strigosum, *c*.4 mm. longum, *c*.9-ovulatum, stylo e base curvato, apice attenuato, *c*.2 mm. longo, stigmate globoso, 0·2 mm. diametro. *Legumen* ± cylindricum, *c*.7–8-spermum, appresse strigulosum [9]–11–[18] mm. longum, *c*.1·5 mm. latum, endocarpio distincte maculato. *Semina* ± cuboidea, brunnea.

UGANDA, 1921, *C. H. Lankaster* s.n.; S. RHODESIA, Lomagundi distr. Trelawney tobacco research sta., 1200 m., erect, suffruticose, flowers salmon pink, 27.3.1943, *R. W. Jack* 70 K, SRGH 96, K, SRGH; Mtoka distr. on sand, 28.1.1941, *Hopkins* in SRGH 7899 SRGH; Gwelo, Feb. 1931, a bush, *Eyles* 6636 holotype, Makuiso, 1290 m., 90–115 cm. tall, Mar. 1918, *Eyles* 1354 K, SRGH, Marandellas distr., grassland 1500 m., *c*.60 cm., flowers red, 6.1.1949, *Corby* 347 K, SRGH; 4.12.1945, *Rattray* 325 SRGH; 4.2.1943, *Dehn* 727 SRGH; 26.3.1942, *Dehn* 155 SRGH, Makais distr. Rusape, 1350 m., April 1953, *Dehn* in SRGH 43085 K. Umtali distr. Osawazi river valley *Teague* 48; occasional in sand veldt on commonage 1050 m., *c*.60 cm. tall, 29.12.1946, *Fisher* 1184, K, SRGH, Bulawayo distr., Jan. 1925, *Steedman* 112 SRGH, Chilimanzi distr. Umvuma, Mtao *c*.1380 m., 28.10.1931, *Brain* 6333 SRGH; Shacha river, shrub *c*.80 cm. tall under dense cover on Kopje, *Seward* 19/51 SRGH; Victoria *Monro* 1705 holotype of *I. grata* var. *longeracemosa*.

This species differs from *I. grata* E. Mey in its longer racemes, fruits free from glandular hairs, shorter calyx lobes and leaves usually longer, with the leaflets wider in relation to their length.

C.3.i.7. **I. phyllanthoides** *Bak.*, F.T.A. **2**, 96 (1871); L.T.A. No. 121.

ANGOLA, Huilla Nene-Empalanca, 1500 m., *Welwitsch* 2044 type.

The fruits are unknown. This species has been confused with *I. dendroides* Jacq. which is a stiffly erect annual herb with pedicels over 1 mm. long in flower and over 2 mm. in fruit whereas *I. phyllanthoides* is a woody suffrutex with pedicels not above 1 mm. long in flower.

C.3.i.8. **I. vicioides** *Jaub. & Spach*, Ill. Pl. Or. t. 481 (1856); L.T.A. No. 99; F.C.B. **5**, 169.

α. var. **vicioides..**

I. tenuicaulis Klotzsch, Pet. Moss. Bot. 50 (1862).

I. transvaalensis Bak.f., Vierteljahrschr. Nat. Ges. Zur. **49**, 193 (1904).

I. divaricata De Wild., Bull. J. Bot. Et Brux. **8**, 149 (1923) non Jacq.

I. semlikiensis Robyns & Boutique, Fl. Parc. Nat. Alb. **1**, 299 (1948).

B. CONGO, East, *Bequaert* 5245 BR holotype of *I. divaricata* De Wild and *I. semlikiensis*; ERITREA, *Pappi* 1711 BM; ETHIOPIA (Dscha Dscha July 1853, *Schimper* P holotype); KENYA, K3–4; TANGANYIKA, T4; PORT. E. AFR. (*Peters* s.n. B† holotype of I. *tenuicaulis*); S. RHODESIA; TROP. BECHUANALAND; S.W. AFRICA; EXTRA TROP. BECHUANALAND; TRANSVAAL, *Schlechter* 4176 isotype of *transvaalensis*.

β. var. **rogersii** (*R. E. Fries*) *Gillett* stat. nov.

I. rogersii R. E. Fries, Wiss. Ergeb. Schwed. Rhod. Kongo Exp. 1911–12 **1**, 80 (1914); L.T.A. No. 196.

I. cognata N.E. Br., Kew Bull. **1925**, 147.

I. hislopii Bak.f., L.T.A. 146 No. 175 (1926).

B. CONGO, Sakania, *Rogers* 10039; KENYA, K4; TANGANYIKA, T1–2; PORT. E. AFR., *Garcia in Mendonça* 382 K, LISC; NYASALAND; N. RHODESIA (*R. E. Fries*

235 UPS holotype), *Rogers* 3496, etc.; S. RHODESIA, *Hislop* 2 holotype of *I. hislopii*, etc.; TRANSVAAL, *Rogers* 23929 holotype of *I. cognata*; NATAL, *Gerrard* 1074.

γ. var. **occidentalis** *Tiss.*, Bull. Mus. Hist. Nat. Par. 2nd Ser. **3,** 165 (1931). OUBANGUI CHARI, *Le Testu* 3661 P, 3792 P syntypes.

Both vars. α and β vary considerably. They may flower either as annuals or as perennials. In perennial plants there is a great difference in appearance between shoots flowering close to the ground after a fire and those which have grown out fully. The type of *I. hislopii* is a specimen of the latter kind. Extreme forms of var. β are very different from var. α but intermediates are so numerous that they must be united in one species. From his statement "poils lâches ou apprimés" it seems that Dr. Cronquist treats var. *rogersii* as a form of *I. vicioides*. He does not however deal with the name, presumably, because both Fries and E. G. Baker wrongly treat Sakania as in N. Rhodesia. *Exell. Mendonça and Wild* 1500 BM from the Matopos, S. Rhodesia may represent an undescribed species allied to *I. vicioides*. The leaflets are *c*.4 mm. long and the pods, which are somewhat curved, *c*.5 mm. long.

C.3.i.9. **I. livingstoniana** *Gillett* sp. nov.

Species propter carinam rostratam vexillo longiorem insignis.

Suffrutex ramis sublignosis, pilis brevibus subappressis subsparse vestitis. *Folia* 3-, rarius 2- vel 4-jugata; stipulae deltoideo-subulatae, *c*.1·5 mm. longae; rhachis sparse appresse pilosa, *c*.3 cm. longa, petiolo 9–12 mm. longo incluso, ultra foliola lateralia 1–3 mm. prolongata; petioluli brunnei 1 mm. longi; foliola aequalia plus minusve elliptica, undique pilis medifixis appressis subsparse conspersa, costa mediana superne impressa inferne prominente. *Racemi c*.12-flori, ad 4 cm. longi, pedunculo *c*.5 mm. longo incluso, rhachide sparse appresse pilosa, bracteis lanceolatis, 1–2 mm. longis, deciduis, pedicellis 0·5 mm. longis. *Calyx* appresse pilosus, tubo 0·5 mm. longo, lobis subulatis, aequalibus, *c*.2 mm. longis. *Vexillum* orbiculare, extus patente pubescens, 3·5 mm. longum, apiculo 0·2 mm. longo incluso, 3·7 mm. latum. *Alae* glabrae, 0·3 mm. unguiculatae, in tota 3·7 mm. longae, lamina basin 1 mm., apicem versus 1·7 mm., lata, apice acuta. *Carina* ad comissuram marginemque pubescens, rostrata, calcare lato obtuso 0·5 mm. longo, in tota 4·5 mm. longa. *Stamina* omnia fertilia, filamento libero 2 mm. longo, connatis 2·5–3 mm. longis, antheris ellipticis, basin muticis, in tota 0·5 mm. longis, apiculo fusco 0·1 mm. longo incluso. *Ovarium* strigosum 2 mm. longum, *c*.7-ovulatum, stylo glabro, 1·2 mm. longo, anguste truncato-conico, paulo supra basin per angulum 70° curvato, stigmate depresse globoso, 0·3 mm. diametro. *Legumen* cylindricum *c*.2-spermum, endocarpio valde maculato, 10–12 mm. longum, 2·5 mm. latum, 2 mm. crassum.

N. RHODESIA, Southern Prov. Kabulamwanda 112 km. N. of Choma, 100 m., sandy waste ground, shrubby, forming a round woody clump 70 cm. tall, flowers bright carmine pink, 13.2.1955, *E. A. Robinson* 1100 holotype. Siamambo Forest Res. nr. Choma, dry sandy loam in Dambo, suffrutex with crimson flowers, 2 plants only seen, 31.7.1952, *A. Angus* 101. Dambwa Forest Res. 6 km. N. of Livingstone on Kalahari sands, suffrutex 20–30 cm. tall, in fruit only, 10.1.1952, *F. White* 1855, Barotseland, Sesheke distr. *Miss A. E. Gairdner* 215, 221.

The species varies somewhat: the other specimens have longer and more copious hairs than the type. In *Gairdner* 221 the keel is shorter than that of the type (4·1 instead of 4·5 mm.). In both the specimens which show fruit the pods appear normal but, on opening, are found to hold very few seeds.

C.3.i.10. **I. heterocarpa** *Welw. ex Bak.*, F.T.A. **2,** 90 (1871); L.T.A. No. 132.

ANGOLA, Huilla, *Welwitsch* 2051 K, BM type.

A little understood species which might be but one more form of *I. hilaris* Eckl. & Zeyh. It is doubtful whether the heterocarpy seen in the type is normal. A kind of fortuitous heterocarpy due to the failure of some ovules to develop (perhaps because of an insufficiency of pollen?) is fairly common in *Indigofera* and may well be all that is involved here.

C.3.i.11. **I. hermannioides** *Gillett* sp. nov.

Herba perennis ad 40 cm. alta, caulibus e radice lignosa robustis, $\pm$ complanatis, rubescentibus, fulvo-hispidis, pilis 1·5 mm. longis, erectis, patentibus vel subprocumbentibus. *Folia* 1-foliolata; stipulae hispidae, e base *c*.1 mm. lata attenuatae, ad 1cm. longae; petiolus hispidus 2–7 mm. longus; petiolulus *c*. 1 mm. longus; stipellae caducae filiformes 2–3 mm. longae; lamina ovato-oblonga vel oblongo-elliptica, basin late cuneata, apice $\pm$ rotundata, costa nervisque primariis utrinque 8–9 superne impressis inferne prominentibus fulvo-hispidis, utrinque albostrigosa, ad 7 cm. longa et 2·6 cm. lata. *Racemi* axillares fulvo-hispidi, foliis breviores, ad 3 cm. longi, pedunculo ad 6 mm. longo incluso; bracteae anguste lanceolatae, hispidae, caducae, ad 5 mm. longae; pedicelli 1–1·5 mm. longi, post anthesin patentes. *Calyx* fulvo-hispidus, pilis ad 0·8 mm. longis, tubo 0·8 mm. longo, lobis aequalibus basin triangulis longe attenuatis, *c*.4 mm. longis. *Vexillum* extra albo strigosum, ellipticum, 4·3 mm. longum, 3·5 mm. latum. *Alae* apice pubescentes, basin 0·4 mm. unguiculatae, suboblongae, 4 mm. longae, 1·2 mm. latae. *Carina* ad margines pilosa, erostrata, laterale 1 mm. calcarata, 4 mm. longa. *Stamina* omnia fertilia, filamentis *c*.4 mm. longis, antheris basin muticis *c*.0·5 mm. longis, mucrone apicale 0·1 mm. longo incluso. *Ovarium* *c*.6-ovulatum, 2 mm. longum; stylus attenuatus, curvatus, glaber, 1·5 mm. longus; stigma globosum, 0·2 mm. diametro. *Legumen* paulo immaturum rectum, crispe et dense albo-vel fulvo-strigosum, pilis *c*.0·2 mm. longis, *c*.4-spermum, ad 11 mm. longum, 1·9 mm. latum, 2·3 mm. crassum, sutura 0·7 mm. lata, rostro apicale (styli basi persistente) paulo deflexo, endocarpio maculato. Semina matura haud visa.

Tanganyika, Songea distr.; *c*.29 km. ENE. of Songea, 1050 m., in *Brachystegia-Uapaca* woodland on red sandy soil; standard pale whitish pink outside, dull cerise inside; wings bright cerise; keel whitish below, cerise at apex and along upper margin, 26.3.1956, *Milne-Redhead and Taylor* 9342 type. 1 km. S. of Gumbiro, *c*.50 km. N. of Songea, 900 m., in *Brachystegia-Uapaca* woodland on sand, 9.5.1956 (unripe fruit) *Milne-Redhead and Taylor* 10126.

This species resembles *I. erythrogramma* Welw. ex Bak. but differs in its larger leaves, longer petioles, narrower stipules, more numerous flowers and spreading, thicker, fewer-seeded fruits.

C.3.i.12. **I. taylori** *Gillett* sp. nov.

Herba perennis, radice longa, sublignosa, caulibus prostratis, paulo complanatis, fulvo-hispidis, pilis ad 1·5 mm. longis. *Folia* 1-foliolata; stipulae hispidae, e basi 1 mm. lata sensim attenuatae, ad 7 mm. longae; petiolus hispidus, 1–3 mm. longus; petiolulus 1–2 mm. longus; stipellae haud visae; lamina elliptica basin cordata, truncata vel late cuneata, apice rotundata, utrinque crispe albostrigosa, costa nervisque primariis utrinque 7-8 superne impressis, subtus prominentibus, fulvo-hispidis, ad 4 cm. longa et 2.7 cm. lata. *Racemi* axillares, tandem foliis longiores, fulvo-hispidi, rariter ramosi, ad 14 cm. longi, pedunculo ad 4 cm. longo incluso, sed saepius multo breviores; bracteae anguste lanceolatae ad 5 mm. longae; pedicelli 1–1·5 mm. longis, post anthesin patentes. *Calyx* fulvo-hispidus, pilis ad 0·8 mm. longis, tubo 0·8 mm. longo, lobis aequalibus, anguste triangularibus, $\pm$ attenuatis, 3–4 mm. longis. *Vexillum* extra albo-strigosum, ovale, 4 mm. longum, 3 mm. latum. *Alae* glabrae, 0·3 mm. unguiculatae, lamina $\pm$ oblonga, 3·5 mm. longa, ad 1·5 mm. lata. *Carina* marginibus apiceque albo piloso, erostrata, laterale 1 mm. calcarata, 4 mm. longa. *Stamina* omnia fertilia, filamentis 3–4 mm. longis, antheris basin muticis, in toto 0·5 mm. longis, mucrone apicale 0·1 mm. longo incluso. *Ovarium* pilosum *c*.6-ovulatum, *c*.2 mm. longum, stylo glabro curvato *c*.2 mm. longo, stigmate globoso, 0·2 mm. diametro. *Legumen* (paulo immaturum) rectum, teres, dense crispe strigulosum, pilis ad 0·3 mm. longis, *c*.5-spermum, ad 11 mm. longum, *c*.2 mm. latum et crassum.

TANGANYIKA, Songea distr. E. of River Luhimba *c*.28 km. N. of Songea 1020 m.; *Brachystegia-Uapaca* woodland on grey sandy soil; perennial with long cylindrical roots, shoots radiating in all directions lying flat on the ground, standard salmon pink inside; wings cerise; keel whitish, 29.1.1956, *Milne-Redhead and Taylor* 8572 type. *c*.1·5 km. W. of River Luhimba just W. of Songea-Gumbiro Rd., 990 m., *Brachystegia-Uapaca* woodland on grey sandy soil, 20.3.1956 (in young fruit) *Milne-Redhead and Taylor* 8572A. Near Mshangano fishponds by River Luhira N. of Songea, 990 m., secondary *Brachystegia-Uapaca* woodland on grey sandy ground, 5.5.1956 (in fruit), *Milne-Redhead and Taylor* 8572B.

Closely related to *I. hermannioides*, from which it differs in its completely prostrate habit, the shorter leaves, broader at the base, shorter petioles and longer inflorescences. The difference in month of flowering and in habitat (grey instead of red sand) may be significant. Mr. Milne-Redhead informs me that both species seem to be very local in the Songea district and were not seen except at the spots where they were collected, where however they were common. *I. taylori* seems to set very little seed, almost all flowers falling off.

Subsect. C.3.j. **Atratae** subsect. nov.

Type species *I. atriceps* Hook.f.

Centre of distribution S. Central Africa.

Folia simplicia vel pinnata, rhachide ultra foliola lateralia saepius prolongata (*I. melanadenia* interdum excepta). Racemus saepius condensatus. Pili multicellulares glandulosi interdum adsunt. Pili biramosi calycis inflorescentiaeque atri vel fusci vel si albi tum pili atri in axillis foliolorum conspicue glomerati. Calyx mediam staminum saepe excedens, dentibus tubo multo longioribus. Legumina patentia, haud glabra, recta saepius in ratione latitudinis brevia. Endocarpium saepius maculatum. Semina juxtaposita.

A confused and difficult group.

Leaves simple; multicellular hairs present on fruits, stems and leaves:
  Inflorescence, both in flower and fruit, short (not over 2 cm. long apart from
    peduncle); not secund, fruiting pedicels *c*.3 mm. long; stamens 3–4 mm. long
      1. *trachyphylla.*

  Infructescence 2–8 cm. long; fruiting pedicels *c*.2 mm. long:
    Stamens 3–4 mm. long; calyx about half as long; leaves ± tapering at the base;
      inflorescence secund . . . . 2. *johnstonii.*
    Stamens *c*.6 mm. long; calyx *c*.5 mm. long; leaves ± abruptly truncate or
      subcordate at the base . . . . 3. *fuscosetosa.*

Leaves pinnate:
  Hairs on calyx all, or nearly all, black or dark brown; tufts of dark glandular hairs
    in axils of leaflets absent; or inconspicuous; endocarp (where seen) rather
    obscurely spotted:
  Seeds usually more than 3; fruit usually with glandular multicellular hairs:
    Leaflets 5–15; petiole up to 10 mm. or more long; fruit *c*.2 mm. wide . . . .
      4. *atriceps.*

    Filaments *c*.30 per cent or more longer than calyx; biramous hairs mostly
      appressed:

Multicellular hairs on fruit often rather long (0·5–1·0 mm.), conspicuously paler than the rest of the fruit; infructescence rarely more than twice as long as peduncle:

Filaments 5 mm. or more long; fruit usually over 8 mm. long:

Multicellular hairs on fruit only, absent from stems; petiole usually under 3 mm. long . . . . α. ssp. *atriceps.*

Multicellular hairs copious on stems as well as on fruits, most often short, sometimes long, petiole short or long . . . .
β. ssp. *alboglandulosa.*

Filaments under 5 mm. long; fruit rarely over 8 mm. long; petiole usually over 5 mm. long, often over 10 mm.; multicellular hairs on stem often very long (up to 3 mm. or more) . . . .
γ. ssp. *setosissima.*

Multicellular hairs on fruit not over 0·3 mm. long, of the same brown colour as the rest of the fruit, infructescence usually more than twice as long as peduncle; filaments under 5 mm. long:

Petiole over 10 mm., often over 20 mm. long; stems, leaf rhachis, peduncle and fruit densely covered with short glandular hairs (not above 0·6 mm. on the stems) infructescence often over 7 cm. long . . . . δ. ssp. *glandulosissima.*

Petiole usually under 10, often under 5, rarely up to 13 mm. long, glandular hairs on stems, fruits, etc., sparse, but some hairs on stems may be up to 1·5 mm. long; infructescence rarely as much as 5 cm. long . . . . ε. ssp. *rhodesiaca.*

Filaments less than 30 per cent longer than calyx; medifixed hairs copious, those on leaves and stems largely crisped and spreading at the tips; infructescence less than twice as long as peduncle; fruit usually short; petiole short (rarely over 2 mm.); multicellular hairs on stems rarely as much as 1·5 mm. long . . . . θ. ssp. *kaessneri.*

Leaflets 3–7; petiole under 4 mm. long; fruit *c.*1·5 mm. wide; multicellular hairs copious on stems . . . . 5. *ramosa.*

Seeds 1–3; fruit without glandular hairs:

Keel rostrate; standard very pointed; calyx *c.* half as long as the 3 mm. long stamens . . . . 6. *schliebenii.*

Keel not rostrate; standard not pointed; or, where corolla unknown (in No. 8) then calyx nearly as long as androecium:

Infructescence up to 1 cm. long; calyx *c.* half as long as the 4 mm. long stamens . . . . 7. *atricephala.*

Infructescence over 2 cm. long:

Calyx nearly as long as stamens; petiole up to 12 mm. long . . . . . .
8. *dasyantha.*

Plant free from glandular multicellular hairs, or with a few on stems and fruits only:

Filaments *c.*4 mm. long; calyx more than half as long; petiole up to 12 mm. long . . . . α. var. *dasyantha.*

Filaments *c.*3 mm. long; calyx less than half as long; petiole up to 3 mm. long . . . . β. var. *brevior.*

Glandular hairs abundant on stems and leaves; filaments *c.*2·5 mm. long . . . . γ. var. *viscidior.*

Calyx *c.* half as long as stamens; petiole 1–2 cm. long . . . 9. *glaucifolia.*

Hairs on calyx partly or wholly white; tufts of dark glandular hairs in axils of leaflets usually conspicuous; fruit usually without glandular hairs:

Stamens less than 5 mm. long; endocarp strongly spotted:

Fruits densely pilose; leaflets ± appressed-pilose above:

Young stems and leaf rhachis strigose or pubescent; the epidermis clearly visible between the hairs; leaflets 5–9 . . . . . . 10. *setiflora.*

Young stems and leaf rhachis with a dense matted silvery indumentum which hides the epidermis; leaflets 9–13 (–17) . . . . 11. *melanadenia.*

Fruits glabrescent; leaflets 3, glabrous above . . . . 12. *sebungweënsis.*

Stamens 6–10 mm. long; leaflets glabrous above:

Calyx and inflorescence axis with rather sparse appressed hairs; endocarp strongly spotted . . . . 13. *williamsonii.*

Calyx and inflorescence axis white and shaggy with copious long spreading hairs; endocarp obscurely spotted:

Inflorescences borne on the upper parts of erect stems in the axils of 18–19-foliolate leaves . . . . 14. *cecilii.*

Inflorescences, or many of them, borne low down on the stems in the axils of 5–9-foliolate leaves . . . . 15. *sanguinea.*

C.3.j.1. **I. trachyphylla** *Benth. ex Oliv.* in Hook. Ic. Pl. t. 1354 (1881); L.T.A. No. 22; vix sensu Cronquist, F.C.B. **5**, 129 (1954).

TANGANYIKA, T4 *Peter* 38851; PORT. E. AFR., Nr. Vila Cabral, *Pedro & Pedrogao* 3516 EA; NYASALAND, *Buchanan* 65 holotype, *Jackson* 1642; N. RHODESIA, Abercorn distr., *Mrs. Richards* 665, 1476, 4536, 4690, 5671.

C.3.j.2. **I. johnstonii** *Bak.f.*, L.T.A. 105, No. 35 (1926).

"*I. trachyphylla Benth.*" sensu Cronquist, F.C.B. **5**, 129 (1954)? non Benth. ex Oliv.

B. CONGO, Haut Katanga (*Verheyen in De Witte* 3195 BR, 3210 BR e descr.); NYASALAND, *Whyte* s.n. holotype, *Jackson* 1570; N. RHODESIA, W. Prov. Mwini-lunga, *Milne-Redhead* 3842, Unfulira, *Eyles* 8323; Kitwe, *Fanshawe* 818; N. Prov. Mpika, *Fanshawe* 1952.

C.3.j.3. **I. fuscosetosa** *Bak.*, Kew Bull. **1897**, 256; L.T.A. No. 38; F.C.B. **5**, 129.

B. CONGO, South, fide Cronquist; TANGANYIKA, T7 *Stolz* 630; NYASALAND, *Whyte* 290, 374 syntypes.

C.3.j.4. **I. atriceps** *Hook.f.*, J. Linn. Soc. **7**, 190 (1864); L.T.A. No. 183; F.C.B. **5**, 165.

α. ssp. **atriceps.**

I. masukuensis Bak., Kew Bull. **1897,** 256; L.T.A. No. 180.

BR. CAMEROONS, *Mann* 1303, 2024 syntypes, etc.; B. CONGO, fide Cronquist; UGANDA, U1; KENYA, K4 Mt. Kenya 3000 m. *Schelpe* 2646 BM; TANGANYIKA, T3, T7; NYASALAND, *Whyte* s.n. holotype of *I. masukuensis,* etc.

β. ssp. **alboglandulosa** (*Engl.*) *Gillett*, Kew Bull. **1955**, 580 (1956).

I. alboglandulosa Engl., Hochgebirgsflora Afrikas 258 (1892); L.T.A. No. 200.

I.djalonica A. Chev. ex Bak.f., L.T.A. No. 179 (1926); non ex Hutch. & Dalz., F.W.T.A. **1**, 391 (1928), Kew Bull **1929,** 16; nec sensu Bak.f., L.T.A. 879 (1930).

I. setosissima Harms var. *major* Cronquist, Bull. J. Bot Et. Brux. **22**, 226 (1952), F.C.B. **5**, 147.

Fr. Guinea, *Maclaud* 103 P, Fouta Djalon *Chevalier* 12854 P p.p. type of *I. djalonica* Chev. ex Bak.f. (1926); Sierra Leone, *Nicholls* 10; Br. Cameroons, *Maitland* 295, 808, 921, 1215, *Mildbraed* 10831; B. Congo, Eastern Mts., *Lejeune* 99 BR holotype of *I. setosissima* var. *major*; Ethiopia (1863) *Schimper* 1547 isotype; A.-E. Sudan, Imatongs, *Johnston* 1456, *Thomas* 1785; Uganda, U1–2; Kenya, K3, K5; Tanganyika, T2, T5–8; Port. E. Afr., Niassa Prov., *Torre* 104, 163 Pedro & Pedrogao 406, EA; Nyasaland; N. Rhodesia, Nyika plateau, *White* 2555.

γ. ssp. **setosissima** (*Harms*) *Gillett* stat. nov.

I. setosissima Harms in Mildbraed Wiss. Ergeb. Deutsch. Zentr. Afr. Exp. **2**, 252 (1911); L.T.A. No. 201; F.C.B. **5**, 146.

I. elskensii Bak.f. ex De Wild., Pl. Beq. **2**, 555 (1924); L.T.A. No. 181.

B. Congo, East and South-East, *Elskens* 243 BR holotype of *I. elskensii* (*Mildbraed* 577 B† holotype), *Lebrun* 9072; Ethiopia, South-West, *Mooney* 6080; Uganda, U2, U4; Kenya, K3; Tanganyika, T1–3, T7; N. Rhodesia, Abercorn distr.

δ. ssp. **glandulosissima** (*R.E. Fries*) *Gillett* stat. nov.

I. secundiflora Poir. var. *glandulosissima* R. E. Fries, Schwed. Rhod. Kongo Exp. 80 (1914).

I. sousae M. A. Exell, Bull. Soc. Brot. 2nd ser. **12**, 8 (1927).

Tanganyika, T7 *Davies* 186, 614; Port. E. Afr., Niassa Prov., *Sousa* 1284 isotype of *I. sousae*, *Pedro & Pedrogao* 34/34 EA; Nyasaland, *Buchanan* 153, *Smuts* 2114; N. Rhodesia, N. Prov. *Mrs. Richards* 5115, Central Prov., *Hutchinson & Gillett* 3679; W. Prov., *R. E. Fries* 363 UPS holotype; *Fanshawe* 1180.

δ. ssp. **rhodesiaca** *Gillett* ssp. nov.

Ab subspeciebus aliis ut in clavi distinguitur. Ab *I. glaucifolia* Cronquist, etiam affine, fructibus ramisque glandulosis, petiolisque brevioribus differt.

Herba semilignosa *c*.80 cm. alta, ramis rectis rigidis, pilis medifixis appressis albidis, et pilis multicellularibus erectis subpurpureis ad 1·0 [1·5] mm. longis, subsparse indutis. *Folia* 4–6-jugata; stipulae subulatae, 3–4 mm. longae; rhachis pilis medifixis, albidis, plerumque appressis, et pilis paucioribus, brevibus, erectis, glanduligeris, obsita, ultra foliola lateralia 3–5 mm. prolongata, in tota, petiolo 2–4 [–6] mm. longo incluso, 3–4 cm. longa; foliola oblanceolata vel cuneato-elliptica, apiculata, utrinque pilis medifixis albidis 0·4–0·5 mm. longis appressis subsparsis, inferne paucis inter eos subbrunneis superne paucis etiam erectis brevissimis glanduligeris obsita, 8–10 mm. longa, 4–6 mm. lata. *Racemi* multiflori 3–4 [–6] cm. longi, pedunculo 5–15 mm. longo incluso; rhachis pilis medifixis saepe crispidulis plerumque brunneis dense et pilis glanduligeris brevissimis sparse obsita; bracteae lanceolatae, brunneae, *c*.1·5 mm. longae, mox deciduae; pedicelli *c*.1 mm. longi. *Calyx* pilis brunneis dense vestitus, tubo *c*.1 mm. longo, lobis deltoideo-acuminatis *c*.1·5 mm. longis. *Vexillum* ellipticum, extus pilis brunneis medifixis dense obsitum, 4·5 mm. longum, 4 mm. latum. *Alae* glabrae, 3·7 mm. longae. *Carina* extus pilis brunneis appressis dense obsita, haud rostrata, 1·5 mm. anguste calcarata, 4·5 mm. longa. *Stamina* omnia fertilia, filamentis ± 4 mm. longis, antheris ellipticis basin muticis, *c*.0·6 mm. longis, apiculo 0·15 mm. longo incluso. *Ovarium* pilosum *c*.2-ovulatum, *c*.1·5 mm. longum, in stylum 3·5 mm. longum, *c*.1·5 mm. infra stigma globosum abrupte curvatum, sensim attenuatum. [Legumen brunneum, cylindricum, pilis glandulosis, *c*.0·2 mm. longis et pilis medifixis indutum, 1–4-spermum, endocarpio obscurissime maculato, 2 mm. latum, 4–8 mm. longum. *Semina* elliptica nitida, 1·5 mm. longa, 1·0 mm. lata, 0·7 mm. crassa.]

Tanganyika, Songea distr., 3 km. N.E. of Kigonsera, 970 m., April 1956, *Milne-Redhead & Taylor* 9718; ; Nyasaland, Mbawa Experimental Sta. Mzimba 1 m. tall, annual weed, 5.4.1955, *G. Jackson* 1592; N. Rhodesia, Abercorn distr., Kalambo falls, 1200 m., in *Brachystegia polyantha* woods, herb 120 cm. tall not common, 14.5.1936 in fruit, *B. D. Burtt* 6222. Chilongowelo side road, 1440 m., sandy soil at edge of old road, *c.*1 m. tall. flowers red, 30.4.1952, *Mrs. Richards* 1657. Chilongowelo, 1470 m., in woodland, among grass, etc., yellow loam, 80 cm. tall, bushy, flower pink,8.3. 1955, *Mrs. Richards* 4841 holotype, ibid. 10.4.1955 *Mrs. Richards* 5370/A. Copper belt, Ndola, plateau woodland, fibrous herb 1 m. tall, flowers pinkish red, 7.3.1954, *D. B. Fanshawe* 949; ibid. 50 cm. tall, in fruit 10.5.1954 *Fanshawe* 1188. Kitwe, in fruit, 12.5.1955, *Fanshawe* 2278.

This subspecies approaches *I. glaucifolia* and a plant which Dr. Cronquist treats, as a brown haired form of *I. shinyangensis*, e.g. *Quarré* 1091, from Elizabethville and *Mooney* 5893 from South-West Ethiopia.

θ. ssp. **kaessneri** (*Bak.f.*) *Gillett* stat. nov.

I. kaessneri Bak.f., J. Bot **41**, 326 (1903); L.T.A. No. 184.

" *I. secundiflora Poir. var.?* " Vatke, Oest. Bot. Zeitschr. 1879, 221.

I. secundiflora Poir. var. *holstii* Bak.f., J. Bot. **41,** 326 (1903).

I. holstii (Bak.f.) Bak.f., L.T.A. 153, No. 203 (1926).

ETHIOPIA, Mega, *Gillett* 14377; UGANDA, U1–3; KENYA, K3, K4 *Kaessner* 876 holotype, *Hildebrandt* 2796, K7; TANGANYIKA, T1–2, T3 *Holst* 8987 isotype of *I. secundiflora* var. *holstii* and *I. holstii*, T6 Uluguru *Schlieben* 3626 BM; N. RHODESIA, Abercorn distr., *Mrs. Richards* 4982, 5085, 5732.

Among the subspecies here recognised ssp. *alboglandulosa* occupies as it were a central position, intermediates occurring between it and most of the other subspecies. Spp. *atriceps* and *alboglandulosa* occur at higher altitudes than any other tropical African species of Indigofera (up to at least 3500 m.). At lower altitudes some forms of *I. atriceps* approach *I. secundiflora* but may be distinguished as follows::

<table>
<tr><td align="center">*I. atriceps*</td><td align="center">*I. secundiflora*</td></tr>
<tr><td>1. Many of the medifixed hairs on calyx inflorescence and fruit dark brown or black.</td><td>All medifixed hairs white.</td></tr>
<tr><td>2. Fruits nearly always more than twice as long as wide.</td><td>Fruits about twice as long as wide.</td></tr>
<tr><td>3. Glandular hairs, or dark medifixed hairs, more conspicuous than white hairs on fruit.</td><td>White medifixed hairs more conspicuous than other types of hairs on fruit.</td></tr>
<tr><td>4. If glandular hairs abundant on stems, calyx much shorter than filaments.</td><td>Calyx about as long as filaments.</td></tr>
<tr><td>5. If biramous hairs on stems spreading, petiole short (under 3 mm.—ssp. *kaessneri*).</td><td>Petiole over 3 mm. long and biramous hairs on stems spreading.</td></tr>
</table>

C.3.j.5. **I. ramosa** *Cronquist*, Bull. J. Bot. Brux. **22,** 224 (1952); F.C.B. **5,** 142. B. CONGO, *Lejeune* 257 BR holotype; TANGANYIKA, T1 *Haarer* 2206 K, EA; NYASALAND, Zomba mt. *Banda* 207 BM.

This may prove to but a form of *I. atriceps*. The plants cited above differ from each other not less than they do from some forms of that polymorphic species.

C.3.j.5A. *Bullock* 2789.

TANGANYIKA, T4 Ufipa plateau Mbisi 2250 m., among grasses on rocky slope near a summit, suberect, flowers red, 2.4.1950.

Leaflets and floral parts larger than those of *I. atriceps*, covered all over with long glandular hairs. Fruits not seen: may represent an undescribed species.

C.3.j.6. **I. schliebenii** *Harms*, Not. Bot. Gart. Berl. **11**, 399 (1932); F.C.B. **5**, 156.

B. Congo, East, *Lebrun* 3868 BM; Tanganyika, T7, Njombe distr. near Lupembe, 35°10′E, 9°12′S, *Schlieben* (173), 408 BM isotype (480) *Frau Schlieben* 86.

C.3.j.7. **I. atricephala** *Gillett* sp. nov.

I. dasyanthae affinis; ab ea ut in clavi differt.

Herba erecta, ramis pilis appressis saepius albis sed nonnunquam atris ± sparse obtectis. *Folia* 7-foliolata; stipulae subulatae 2–3 mm. longae; rhachis ultra foliola lateralia 2–4 mm. prolongata, in tota ad 20 mm. longa, petiolo 1–2 mm. longo incluso, indumento ut ad ramos; foliola elliptico-lanceolata aequalia, apice subacuta, apiculata, utrinque pilis albidis appressis, *c.*0·5 mm. longis, ± sparse induta, ad 17 mm. longa, 6 mm. lata. *Racemus* 15–30-florus, confertus, fere globosus; axis pilis atratis vestitus, ad 10 mm. longus, post anthesin haud elongatus, pedunculo ad 40 mm. longo multo brevior; bracteae erectae, mox deciduae, lanceolatae, *c.*1 mm. longae, dense atropilosae; pedicelli pilis atratis dense vestiti, *c.*1 mm. longi, post anthesin patentes. *Calyx* pilis rigidis atratis dense vestitus, tubo *c.*1 mm. longo, lobis aequalibus deltoideo-subulatis *c.*1·2 mm. longis. *Vexillum* ovato-oblongum, extus pilis atratis dense vestitum, apice rotundatum, *c.*4 mm. longum, 2·2 mm. latum. *Alae* glabrae, 0·8 mm. latae, 3·5 mm. longae, unguiculo 0·3 mm. longo incluso. *Carina* extus pilis atratis dense vestita 1·2 mm. acute calcarata, haud rostrata, 4·3 mm. longa. *Stamina* omnia fertilia, filamentis 3·5–4 mm. longis, antheris ellipticis 0·7–0·8 mm. longis, appendice basale rubro, crasso, obtuso 0·05 mm. longo et appendice apicale rubro acuto 0·2–0·3 mm. longo inclusis. *Ovarium* strigosum, *c.*3-ovulatum, 1 mm. longum; stylus 4 mm. longus infra mediam strigosus, superne glaber, *c.*1·5 mm. infra stigma depresso-globosum 0·1 mm. diametro, curvatus. [Legumen 2-spermum pilis rigidis appressis saepius atratis obtectum, *c.*4·5 mm. longum, 1·8 mm. latum, 1·3 mm. crassum.]

Tanganyika, T7, Mbeya distr., Usefwa *c.*33°37′E, 8°50′S, 1650 m. herbaceous, common, 23.3.1932, *R. M. Davies* 143 K holo. EA isotype. "Mtns. E. of L. Nyassa" *Rev. W. P. Johnson* s.n. (in fruit) (Johnson travelled extensively, and this locality may be in either Tanganyika, Port. E. Afr. or Nyasaland.)

C.3.j.8. **I. dasyantha** *Bak.f.*, L.T.A. 148, No. 182 (1926).

α. var. **dasyantha.**

Tanganyika, Songea, *Milne-Redhead & Taylor* 9792; Nyasaland, Kondowe-Karonga, July 1896, *A. Whyte* s.n. p.p. holotype; N. Rhodesia, Abercorn distr. 1955, *Mrs. Richards* s.n.; ibid. Chilongowelo, bush by house, 1440 m., red loam, 65 cm. tall. in fruit, 4.5.1955, *Mrs. Richards* 5511.

The type sheet is a mixture, having between two shoots of this species, one of *I. atriceps* ssp. *glandulosissima*. Baker's description is for the most part based on the two outer shoots, but his description of the corolla, which is not shown on the outer shoots, must be drawn from the inner. The petiole is shorter than the basal leaflets, being 4–6 mm. long. The peduncle is 20–35 mm. long and is much shorter than the fruiting raceme itself which is up to 7 cm. long and markedly secund. The lower calyx lobes are 3·5 mm. long, about as long as the filaments.

β. var. **brevior** *Gillett* var. nov.

A. var. *dasyantha* petiolis brevioribus (2–3 mm., haud 3–12 mm.), calyce breviore (*c.*1 mm., haud 2·5–3·5 mm.), filamentis brevioribus (*c.*3 mm., haud *c.*4 mm.) differt. Legumen saepius 1-spermum, rariter 2-spermum. Pili multicellulari ubique desunt.

Tanganyika, Songea dist., 1·5 km. S. of Gumbiro 880 m., *Brachystegia* woodland on gravelly clay soil, short-lived perennial, standard brown with hairs outside, inside greenish white above, pinkish towards base; wings salmon, keel brown with hairs, 29.3.1956, *Milne-Redhead and Taylor* 9372. Tunduru district: just E. of Songea district boundary, 880 m., *Brachystegia-Uapaca* woodland on eroded slope on gravelly soil, short-lived perennial, fruits brown, 5.6.1956, *Milne-Redhead and Taylor* 10605.

γ. var. **viscidior** *Gillett* var. nov.

A var. *breviore* praecipue pilis glandulosis multicellularis ad 0·8 mm. longis ad caules, foliorum racemorumque rhaches, margines nervosque foliolorum numerosissimis, paucis etiam ad paginam superiorem et inferiorem foliolorum, differt. Ad flores tandem hi fili fere desunt. Foliola 5–7, ad 2·5 cm. longa et 1 cm. lata. Petiolus 3–4 mm., calyx 1·5–2 mm., filamenta *c.*2·5 mm. longa. Ovula 2; legumen haud visum.

TANGANYIKA, *c.*6·5 km. W. of Songea, 990 m., sandy ground between *Brachystegia-Uapaca* woodland and boggy grassland, corolla pubescent outside with brown appressed hairs, salmon red within, 6.2.1956, *Milne-Redhead and Taylor* 8621.

C.3.j.9. **I. glaucifolia** *Cronquist*, Bull. J. Bot. Et Brux. **22**, 219 (1952); F.C.B. **5**, 155.

B. CONGO, South, *Ritschard* 1540 BR holotype; N. RHODESIA, Solwezi distr. Mbulungu stream near Mutanda bridge, at edge of evergreen shrubs, 15.7.1930, *Milne-Redhead* 708 (in fruit); Mwinilunga distr. by river Matonchi at edge of evergreen vegetation, in shade, 9.2.1938, *Milne-Redhead* 4512 (in flower); Abercorn distr., *Mrs. Richards* 4868, 5684.

The fruits, which have not hitherto been described, are ± cylindrical, spreading or somewhat ascending, 2–3-seeded, ending in a short subulate point formed by the persistent style base, covered with closely appressed blackish medifixed hairs, 5–8 mm. long, 1·6 mm. wide. The seeds are smooth, shining, dull yellow, rounded-oblong, elliptical in T.S., 1·5 mm. long, 1·1 mm. wide, 0·5 mm. thick. This species is very close to certain forms of *I. atriceps* ssp. *rhodesiaca*.

C.3.j. 10. **I. setiflora** *Bak.*, F.T.A. **2**, 90 (1871); L.T.A. No. 174.

"*I. confusa Prain & Bak.f.*" sensu Bak.f., L.T.A. 143, No. 161 (1926) non Prain & Bak.f., J. Bot. **40**, 143 (1903).

I. accepta N.E. Br., Kew Bull. **1925,** 153.

PORT. E. AFR., Manica and Sofala, *Mendonça* 669 LISC, 3844 LISC, *Torre* 4040 K, LISC *Barbosa in Mendonça* 1199 K, LISC; Lourenco Marques *Torre* 7095 K, LISC; S. RHODESIA, *Eyles* 2038 K, SRGH, etc. Wankie, Bulawayo, Gwelo, Hartley, Salisbury, Marandellas, Inyanga, Makoris, Umtali districts; N. RHODESIA, Mazabuka and Livingstone districts; ANGOLA, *Welwitsch* 2020 type; N. TRANSVAAL, *Rogers* 21995 holotype of *I. accepta*.

The true *I. confusa* Prain & Bak.f. (*I. arrecta* Benth. ex Harv. non Hochst. ex A. Rich.), whose syntypes are *Burke* 340 and *Zeyher* 474 from the Transvaal, differs from all the Rhodesian and Angolan material of *I. setiflora* in, among other things, a leaf rhachis which is not prolonged beyond the last pair of lateral leaflets. It is conspecific with *I. pretoriana* Harms (1898). The greater length of the peduncle which N. E. Brown used to separate the two plants in Burtt Davy, Man. Flora Transvaal **2**, 361–6 is much too variable a character. *I. atrata* N. E. Brown and *I. egens* N. E. Brown are also closely related to *I. pretoriana*, and also have a ± sessile terminal leaflet.

I. setiflora varies greatly. Lax slender forms, almost free from black hairs and with short leaflets, such as *Jack* 199 SRGH, and *Torre* 7095 LISC approach *I. vicioides* var. *rogersii*, from which they are distinguished by their short thick pods. Other robust plants, with dense inflorescences and copious black hairs from the eastern, high-rainfall, areas of Port E. Afr. and S. Rhodesia, *Mendonça* 6169, *Barbosa in Mendonça* 1199 LISC and *Hopkins* in SRGH 7080, 8840 from Umtali, appear at first sight specifically distinct but are connected by numerous intermediates with the typical form of the species. It will probably eventually be useful to delimit infra-specific taxa but this should be postponed till more complete material is available.

C.3.j.11. I. melanadenia *Benth. ex Harv.*, Fl. Cap. **2**, 195 (1862).

S. RHODESIA, Gutu, Chikwanda Reserve, *D. A. Robinson* 307 K. SRGH; Marandellas *Corby* 411 K, SRGH, *Wild* 3285 K, SRGH; *Matopos* Martineau 175 SRGH; EXTRA TROP. BECHUANALAND; TRANSVAAL; ORANGE FREE STATE, *Burke* 237 syntype.

C.3.j.12. I. sebungweënsis *Gillett* sp. nov.

Ab *I. setiflorae* et *I. melanadeniae* foliis 3-foliolatis, leguminibusque glabrescentibus differt.

Suffrutex ramosis valde lignosus *c.*1 m. altus, ramis vetustioribus cortice atrogriseo multo fissurato, ramis junioribus cortice subrubro, pilis minutis (*c.*0·2 mm. longis) appressis, dense vestitis. *Folia* pinnatim 3-foliolata; stipulae minutae, deltoideae, argenteae, 0·5 mm. longae; rhachis petiolulique rubri, minute puberuli, rhachide in tota 6–10 mm. longa, parte ultra foliola lateralia petiolo aequante, petiolis *c.*1·5 mm. longis; pili nigri glandulosi breves in axillis foliorum et foliorum et ad basin petioluli folioli terminalis dense glomerati; foliola aequalia, lanceolata, basin attenuata, superne glabra, inferne minute appresse pilosa, *c.*14 mm. longa, 4 mm. lata. *Racemi* multiflori, 2–5 cm. longi, pedunculo 2–5 mm. longo incluso; rhachis minute appresse pilosa; bracteae breves caducae; pedicelli *c.*1 mm. longi, post anthesin erecto-patentibus, pilis glandulosis atro-rubris in axillis dense glomeratis. *Calyx* minute appresse puberulus, tubo 0·5 mm. longo, lobis deltoideis inaequalibus, infimo 1·5 mm. longo, lateralibus 1 mm. longis, superioribus 0·7 mm. longis. *Corolla* haud visa. *Filamenta* 2 mm. longa. *Ovarium* appresse pilosum, stylo *c.*2 mm. longo, cylindrico, paulo supra mediam abrupte curvato. *Legumen* glabrescens subcylindricum, 2-spermum, 7 mm. longum, in plano commisurae 2·5 mm. latum, 3 mm. crassum, commissura inferiore 0·5 mm. lata, endocarpio praecipue in 4 lineis rubro-maculato. *Semina* grisea, paulo foveolata, subcuboidea, 1·9 mm. longa, 1·5 mm. lata, 1·3 mm. crassa.

S. RHODESIA, Sebungwe distr., Kariangwe hill. *c.*27°32′E, 17°59′S, on sandstone outcrops, small shrub 1 m. tall. mTonga name mWangabwi, in fruit, 2.7.1951, *D. F. Lovemore* 80 (SRGH 33429) K holo SRGH isotype. Same locality, in fruit, 12.8.1951, *J. A. Whellan* 530 (SRGH 34006) K. Susulu, on shallow soil over rocks, small shrub 1 m. tall, in fruit 28.5.1947, *O. West* 2339 (SRGH 25433) SRGH.

C.3.j.13. I. williamsonii (*Harv.*) *N. E. Brown* in Burtt Davy. Man. Flor. Transvaal **1**, 45 (1926).

I. eriocarpa E. Mey var. *williamsonii* Harv., Fl. Cap. **2**, 194 (1862).

S. RHODESIA, Gwelo distr. Lalapanzi, *Walters* 2432; Lomagundi distr. Trelawney, *Jack* 230 K, SRGH; NATAL, Port Natal Peddie 1840 *Harvey* s.n. isotype *Medley Wood* 258, etc.; E. CAPE PROV., *Tyson* 2742.

C.3.j.14. I. cecilii *N. E. Br.*, Kew Bull. **1906**, 101.

"*I. rehmannii Bak.f.*" sensu Bak.f., L.T.A. 161, No. 242 (1926) p.p. non Bak.f., Vierteljahrschr. Nat. Ges. Zur. **49**, 190 (1904).

PORT. E. AFR., Chirinda, *Swynnerton* 262 BM, 263; Namaacha on Swazi frontier 600 m. *Gomez e Souza* 438; S. RHODESIA, Inyanga distr., *Cecil* 186 holotype, *Eyles* 8508 K, SRGH, *Hopkins* in S.R.G.H. 7855, 7874. Melsetter distr., *A. O. Crook* M81, K, SRGH, *Sturgeon* in SRGH 30656 K, SRGH, *Wild* 4568, *Swynnerton* 1460, Chipinga distr., *Chase* 502, *Rattray* 1134.

E. G. Baker identified this species with his *I. rehmannii* from Houtbosch in the Transvaal. However, all the specimens cited above have the leaflets glabrous above, while the type of *I. rehmannii, Rehmann* 6224 Z and *Burtt Davy* 5175 and *Bruce & Kies* from the same area have the leaflets pilose above. This single character may seem insufficient ground for retaining *I. cecilii* as a distinct species, but if such narrow distinctions are to be disregarded a whole plexus of related South African species including *I. sanguinea* N. E. Br., *I. comosa* N. E. Br., *I. mollicoma* N. E. Br., *I. oxytropis* Benth. ex Harv., *I. eriocarpa* E. Mey and *I. velutina* E. Mey must be thrown into doubt. On the whole it seems better to retain, in this instance, Mr. N. E. Brown's admittedly extremely fine specific distinction until some South African botanist undertakes a thorough revision of this whole difficult group of species.

C.3.j.15. **I. sanguinea** *N. E. Br.* in Burtt Davy, Man. Flor. Transvaal **2**, XXIV and 370 (1932).

Port. E. Afr., Lourenço Marques, between Namaacha and Impamputa *A. R. Torre* 8000 LISC; Transvaal, *Wilms* 345 holotype, etc.

Subsect. C.3.k. **Psiloceratiae** subsect. nov.

Type species *I. sutherlandioides* Welw. ex Bak.

Centre of distribution South Tropical Africa.

Folia pinnata, rhachide ultra foliola lateralia producta. Calyx quam stamina multo brevior. Pedicellus fructifer suberectus, patens vel reflexus. Ovarium et legumen glabrum vel fere glabrum. Endocarpium saepius maculatum. Semina 4 vel plura saepius inter se bene separata.

Perhaps an unnatural group; most species seem akin to the **Tinctoriae** but *I. omissa* is clearly related to *I. heudelotii* (**Dissitiflorae**) and *I. laxeracemosa* to *I. vicioides* (**Centrae**). Glabrous fruits, outside this section, are found in certain species of the **Dissitiflorae**, in *I. thomsonii* (sometimes) and in one or two species of subgen. *Microcharis*.

I. macrantha, I. macrophylla and *I. mearnsii* have been wrongly described as having glabrous fruits.

I. kirilowii Max. & Palibin, *I. decora* Lindl. and perhaps other spp. from China also belong to this subsection.

Stamens not above 4 mm. long:
Pods slightly torulose, *c.*2·3 mm. wide, abruptly contracted at the apex; leaves subtending inflorescences 5–11-foliolate; endocarp not spotted . . . .
. . . . 1. *laxeracemosa.*
Pods not torulose, *c.*1·4 mm. wide, tapering to a point; leaves subtending inflorescences 3–5-foliolate; endocarp spotted . . . . 2. *omissa.*
Stamens over 6 mm. long:
Stems ± herbaceous, almost glabrous, spreading, often rooting at the nodes; leaflets 5–13, glabrous above, hairs elsewhere sparse and appressed; fruit somewhat curved upwards near apex . . . . 3. *longemucronata.*
Stems usually woody, ± erect, not rooting at the nodes; leaflets rarely glabrous above; fruit (where seen) ± straight:
Standard glabrous outside in the bud, except at the tip and margins; filaments 12–14 mm. long . . . . 4. *baumiana.*
Standard in the bud pubescent outside:
Leaflets usually more than 7, not glabrous above, leaf rhachis pubescent:
Keel about half as long again as standard . . . . 5. *longiflora.*

Keel about as long as, or shorter than standard:
  Stem not triquetrous:
    Leaflets more than 1 cm. long:
      Standard pointed, keel $\pm$ rostrate (lower edge meeting the upper at much less than a right angle) . . . . 6. *fulgens.*
      Stamens *c.*14 mm. long . . . . α. ssp *fulgens.*
      Stamens *c.*10 mm. long . . . . β. ssp *brachybotrys.*
      Standard not pointed, keel rounded or truncate at the apex:
        Calyx teeth much shorter than the tube, triangular; filaments 9–11 mm. long . . . . 7. *sutherlandioides.*
        Calyx teeth as long as the tube, subulate; filaments 6–8 mm. long 8. *lyallii.*
        Hairs on leaf surface dense, soft, weakly spreading . . . . . . α. ssp *lyallii.*
        Hairs on leaf surface stiff, sparse appressed . . . β. ssp *nyassica.*
    Leaflets less than 1 cm. long; filaments 9–11 mm. long 9. *gloriosa.*
  Stem triquetrous, hairs on leaf surface stiff, sparse, appressed . . . . . 10. *homblei.*
Leaflets 5–7, often glabrous above, leaf rhachis $\pm$ glabrous 11. *paucistrigosa.*

C.3.k.1. I. laxeracemosa, *Bak.f.*, J. Bot. **41,** 241 (1903); L.T.A. No. 113.

TANGANYIKA, T3 *Peter* K550; ZANZIBAR, Vaughan 1130, 1422, 1426; PORT. E. AFR., *Speke* 14, *Scott* s.n., *Kirk* s.n. syntypes, etc.; TRANSVAAL, Lowveldt.

This species appears to be restricted to low altitudes. It is very close to certain forms of *I. vicioides*, a species found at higher altitudes, intermediates occur occasionally: e.g. *Codd* 4619, 5416 and *Van der Schyff* 2778 from the Kruger national park.

C.3.k.2. I. omissa *Gillett*, Kew Bull. **1955,** 580 (1956).

FR. SUDAN; DAHOMEY.

C.3.k.3. I. longemucronata *Bak.f.*, J. Bot. **41,** 330 (1903); L.T.A. No. 232.

KENYA, K7 Giryama and Tsimba Mts. 1887 *W. E. Taylor* s.n. BM. holotype, etc.; TANGANYIKA, T3 *Holst* 2995, etc.

C.3.k.4. I. baumiana *Harms* in Warburg and Baum Kunene-Zambesi Exp. 257 (1903); L.T.A. No. 238, excluding all varieties.

B. CONGO, South, Dilolo *Young* 213 BM; N. RHODESIA, West, *Agr. Dept.* 19, *Milne-Redhead* 1101, 2622, 3898; ANGOLA, *Baum* 497 isotype, *Young* 699 BM, 720 BM, 1324 BM, *Carisso and Mendonça* 466 BM, *Gossweiler* 3051 BM, 3272 BM, 2757 BM.

This species has been much confused with *I. sutherlandioides*. The longer pedicels and stamens, the pointed buds and standard, the latter glabrous outside but for the margin and tip, and, perhaps, more strongly reflexed fruits distinguish it. It also seems always to flower on annual shoots arising from a woody rootstock, whereas *I. sutherlandioides* is usually a shrub.

C.3.k.5. I. longiflora *Taub.* in Engl. Pflanzenwelt O. Afr. C. 210 (1895); L.T.A. No. 241.

PORT. E. AFR., Mozambique (*Stuhlmann* 871 B† holotype).

This species is obscure, *Busse* 1096 EZ, probably from near the Rovuma river in S.E. Tanganyika, which was identified by the Germans as *I. longiflora*, is a close match of the type of *I. fulgens* from the same area; in particular it has a keel of about the same length as the standard. In the description of *I. longiflora* the keel is said to be half as long again as the standard, and this character alone seems to separate *I. longiflora* from *I. fulgens*. Further collections are necessary, especially from the type areas, to establish or disprove this distinction.

C.3.k.6. **I. fulgens** *Bak.*, F.T.A. **2,** 101 (1871); L.T.A. No. 237.

α. ssp **fulgens.**

Tanganyika, T4 Konongo Namanyere subdistr., 9.12.1921, *Swynnerton* 41 BM, T8 Rovuma river *Kirk* s.n. holotype, *Meller* s.n., *Busse* 1096 EA, *Gillman* 1056 EA, 1452 EA Nachingwea *Anderson* 957 EA; Port. E. Afr., Sul do Save, Hornoine *Exell, Mendonça and Wild* 581 BM.

β. ssp **brachybotrys** (*Bak.*) *Gillett* stat. nov.

I. brachybotrys Bak., J. L. Soc. **25,** 309 (1890).

Madagascar, *Baron* 5365 holotype.

The difference in flower size seems alone to separate sspp. *brachybotrys* and *fulgens*. I cannot follow Cronquist in uniting *I. fulgens* and *I. sutherlandioides*. Besides the difference given in the key there is a difference in aspect which is most striking when a number of specimens are seen.

C.3.k.7. **I. sutherlandioides** *Welw. ex Bak.*, F.T.A. **2,** 101 (1871); L.T.A. No. 235; F.C.B. **5,** 137 pro maj. parte.

I. baumiana var. *paucijuga* R. E. Fries, Schwed. Rhod. Kongo Exp. 81 (1914).

I. baumiana var. *roseo-violacea* De Wild., Bull. J. Bot. Et Brux. **8,** 145 (1923).

I. baumiana var. *hockii* De Wild., l.c. 146 e descr.

I. noldeae Rossberg, Fed. Rep. **38,** 106 (1935) e descr.

B. Congo, South (*Hock* s.n. BR holotype of var. *hockii*), *Homblé* 898 BR holotype of var. *roseo-violacea*, *Quarré* 3609, *Rogers* 10064, *Hirschberg* 224; Tanganyika, T6 Pugu hills, *MacLean* 1; N. Rhodesia, *R. E. Fries* 833 UPS holotype of var. *paucijuga, Hutchinson & Gillett* 3806, *Fanshawe* 495, *Cruse* 60, 235, *Mrs. Richards* 119, 1558, *Siame* 8, *Angus* 849, *Scott Elliot* 8288, *White* 2697, 3801, *Bullock* 1016, 2136, *Burtt* 6208, *Pole Evans* 2872, *Monks* 22, *Milne-Redhead* 2620, 3662, 3900; Angola, *Welwitsch* 2012 type (*Nolde* 108B† holotype of *I. noldeae*) *Wellman* 1813, *Gossweiler* 9516, 9576.

This species varies greatly, particularly in the shape, number, size and indumentum of the leaflets, but the various characters seem so little co-ordinated that it is hardly possible at this stage to delimit infra-specific taxa. Dr. Cronquist adapts an even wider view of the species than I have done, including within it *I. baumiana* and *I. fulgens*. The fruiting type of *I. dekindtii* Tisserant, (1931) 169, *Antunes* 3104P is not distinguishable from certain forms of this species. On the other hand the flowers, as described by *Tisserant* from *Antunes* 3162 differ greatly in their size.

C.3.k.8. **I. lyallii** *Bak.*, J. L. Soc. **20,** 128 (1883); L.T.A. No. 240.

α. ssp **lyallii.**

"*I. pulchella Roxb.*" sensu Bak.f., L.T.A. 160 non Roxb.

I. obermejerae Bremekamp, Ann. Transvaal Mus. **15,** 242 (1933).

Port. E. Afr., Chimanimani Mts., *Swynnerton* 1458 BM, 1468 K, *Sturgeon and Panton* in SRGH 30697; S. Rhodesia, Melsetter distr., *Chase* 1416 K, SRGH, *Wild* 1947 K, SRGH; Umtali distr., *Eyles* 3638, 6144 SRGH, *Martineau* 324 SRGH; *Mundy* in SRGH 3189, *Swynnerton* 267; Madagascar, *Lyall* 241 holotype,

Baron 761, 969, 3375, 3966, 5039, 5199, *Hildebrandt* 3881, etc.; N. TRANSVAAL, Zoutpansberg, *Obermeyer* 1112 syntype of *I. obermejerae, Gerstner* 5989, *Codd* 8334, K, EA, (PRE).

β. ssp **nyassica** *Gillett* ssp. nov.

A ssp *lyallii* pilis foliolorum rectis, appressis, sparsis differt.

PORT. E. AFR., near Gurue *c*.37°E, 15°30'S, 23.6.1942, *A. J. W. Hornby* 3763; ibid. 1000 m. mountain slopes and banks of streams, shrub 1–3 m. tall *Torre* 5109 K, LISC; NYASALAND, Mt. Mlanje, 1891, *A. Whyte* s.n. holotype; 1300–2100 m., Mar. 1897, *G. Adamson* 359; Lichenya plateau 1860 m., locally dominant with *Polygala, Helichrysum hochstetteri, Myrsine africana, Rubiaceae, Philippia, Dodonaea,* in *Widdringtonia* forest patch on poor rocky bauxite soil on steep slope, crimson flowered much branched shrub up to 2 m. tall, 17.10.1941, *Greenway* 6325 K, EA; 1900 m., tree or shrub up to 6 m., diameter at breast height 15 cm., common in secondary growth, 7.7.1946, (in fruit) *Brass* 16705.

Note (1) E. G. Baker's statement that "*I. pulchella*", *Swynnerton* 267 has "glabrous flowers" is incorrect. (2) From Mr. Brass's field notes this seems to be the tallest and stoutest *Indigofera* species of which I have seen any record.

C.3.k.9. **I. gloriosa** *Cronquist,* Bull. J. Bot. Et Brux. **22,** 219 (1952); F.C.B. **5,** 134.

B. CONGO, South, *De Witte* 41 BR holo. BM isotype.

C.3.k.10. **I. homblei** *Bak.f. & Martin,* Bull. J. Bot. Et Brux. **8,** 151 (1923); L.T.A. No. 228. F.C.B. **5,** 139.

B. CONGO, *Homblé* 1135 BR holotype; UGANDA, U2 *Mainwaring* 3170 EA, *Purseglove* 801, *Bagshawe* 443 BM; KENYA, K3 *Jack* 327, *Bogdan* 242, *Brodhurst Hill* 566, *Bally* 5085; TANGANYIKA, T7 *Davies* 46, 137, 144; NYASALAND, *Whyte* s.n.; ANGOLA, *Gossweiler* 2857 BM, *Nolde* 371 b BM; TRANSVAAL, *Mogg* in PRE 13682, 14723.

As Dr. Cronquist remarks, there is great variation in the length of the bracts, which are sometimes much longer than the flower buds, sometimes inconspicuous. They are usually deflexed at the tip.

C.3.k.11. **I. paucistrigosa** *Gillett* sp. nov.

Suffrutex, (vel herba sublignosa), ad 130 cm. altus, ramis cylindricis brunneis subglabris. *Folia* 5–9-foliolata; stipulae deltoideo-subulatae *c*.2 mm. longae; rhachis fere glabra, superne canaliculata, ultra foliola lateralia 7–12 mm. prolongata, 20–70 mm. longa, petiolo 12–17 mm. longo incluso; stipellae subulatae 1·5 mm. longae; petioluli nigri ad 2·5 mm. longi; foliola obovata vel fere orbicularia, basin paulo attenuata, apice truncata 0·5 mm. apiculata, superne glabra, inferne tenuiter strigosa, strigis appressis subsparsis *c*.0·4 mm. longis, ad 18 mm. longa, 14 mm. lata. *Racemus* multiflorus ad 14 cm. longus, pedunculo 15–30 mm. longo incluso; axis fere glaber; bracteae anguste deltoideae, ad 1·5 mm. longae, mox deciduae; pedicelli appresse pilosuli primo ad 3, deinde ad 4 mm. longi, post anthesin patentes, vel subreflexi. *Calyx* appresse pilosulus inaequalis, tubo superne 1·5 mm. longo, inferne 2 mm. longo, lobis 2 superioribus fere obsoletis, 0·1 mm. longis, 3 inferioribus late deltoideis *c*.0·5 mm. longis, sinu superiore quam alios fere duplo latiore. *Vexillum* extus dense appresse pubescens, lanceolatum, apice subacutum, ad 14 mm. longum, 6·5 mm. latum. *Alae* margine pilosulae, basin incrassatae, suboblongae, apice ± acutae, *c*.11 mm. longae, ad 2·5 mm. latae. *Carina* extus appresse pubescens, *c*.1 mm. obtuse calcarata, subrostrata (margine inferiore ad marginem superiorem ad 45° connivente), ad 15 mm. longa. *Stamina* omnia fertilia, filamentis 13–14·5 mm. longis, antheris oblongis, basin muticis, 1·2 mm. longis, appendice apicale 0·15 mm. longo incluso. *Ovarium* glabrum *c*.12-ovulatum, *c*.13 mm.

longum, versus apicem paulo curvatum, in stylum ascendentem *c*.2 mm. longum sensim attenuatum, stigmate punctiforme. *Legumen* glabrum, versus apicem paulo curvatum, *c*.10–11-spermum, *c*.57 mm. longum, basin *c*.3 mm., superne fere 4 mm., latum, *c*.2 mm. crassum, seminibus inter se distantibus, endocarpio brunneo-maculato.

Tanganyika, T1, Mwanza distr., near Uzuruga camp, Uzinza area, *c*.32°15′E, 2°40′S, 1170 m., stony hill clothed with *Brachystegia microphylla*, 1–1·3 m. tall, flowers salmon red, locally common, 26.3.1937, *B. D. Burtt* 6485 K holo., EA BM isotypes.

A. Carson s.n. from Niomkolo near Lake Tanganyika, N. Rhodesia may represent a form of this species. The flowers are rather smaller (filaments *c*.12 mm. long) and the leaflets appressed-pilose above.

Subsection C.3.l. **Geanthae** subsect. nov.

Type species *I. thomsonii* Bak.f.

Racemi in axillis foliorum reductorum infra folia normalia. Pedicellus fructifer ± patens. Calycis lobi tubo multo longiores.

Corolla pubescent, leaves pinnately 3-foliolate, the rhachis prolonged 9–15 mm. beyond the lateral leaflets; fruit tetragonous; endocarp not spotted . . . .

1. *pseudosubulata*.

Corolla almost glabrous, leaves 1–9 foliolate, the rhachis hardly prolonged beyond the lateral leaflets; fruit not tetragonous; endocarp spotted . . . . 2. *thomsonii*.

C.3.l.1. **I. pseudosubulata** *Bak.f.*, J. Bot. **41**, 264 (1903); L.T.A. No. 169; F.C.B. **5**, 161.

Oubangui, Yalinga, *Le Testu* 2567 P; B. Congo, North-East, fide Cronquist; A.-E. Sudan, Niamniam and Bongoland, *Schweinfurth* 2703, 2994, ser. 2, 77 syntypes.

C.3.l.2. **I. thomsonii** *Bak.f.*, J. Bot. **41**, 189 (1903); L.T.A. No. 26; F.C.B. **5**, 134.

B. Congo, East and South-East; Tanganyika, T4, T7 *Stolz* 2194 UPS, *J. Thomson* s.n. holotype; Nyasaland, North; N. Rhodesia, Abercorn and Mwinilunga districts.

The two specimens, *Milne-Redhead* 1190 and *Angus* 491 from the Mwinilunga district may represent a distinct taxon. They both have leaves commonly 5-foliolate and, in *Angus* 491, up to 9-foliolate. Although both show young fruit (not seen elsewhere) and the leaflets are thus older than usual the leaflets does not reach 2 cm. in length. None of the 7 specimens seen from Tanganyika and Nyasaland, which are all in the early flowering condition, show leaves with more than 3 leaflets and these immature leaflets are up to 4 cm. long. *Bullock* 3307 from Mpanda shows a single old leaflet from the previous year which is over 8 cm. long. It is desirable that further collections showing mature leaves and fruits should be made to establish the nature of the variation within what is here provisionally treated as a single species.

Subsect. C.3.m. **Tinctoriae** (*Bak.*) *Gillett*.

"Group" Tinctoriae Bak., F.T.A. **2**, 68 (1871); L.T.A. 139 emend.

Lectotype species *I. tinctoria* L.

Centre of distribution perhaps in India.

Leaves pinnate, or, less often, trifoliolate. Inflorescence many-flowered, the fruiting pedicels deflexed. Calyx much shorter than corolla, the lobes rarely more than twice as long as the tube, the sinus between the upper two teeth often much wider than the others. Standard densely covered outside with closely appressed ±glistening, frequently brownish hairs; sometimes subpersistent. Fruit 4- or more-seeded, rarely 1–3-seeded, sometimes ± curved. Seeds usually spaced well apart, the endocarp, which may or may not be spotted, forming a thick septum between them.

Stems not creeping and rooting at the nodes:
  Stipules not foliaceous, not over 2 mm. wide:
    Calyx lobes less than twice as long as tube, or, if twice as long filaments under 5 mm. long:
      Pod with a stipe *c.*3 mm. long, or more, dark haired; endocarp spotted:
        Fruit not torulose, stamens *c.*12 mm. long . . . . .1. *podocarpa.*
        Fruit torulose, stamens *c.*5 mm. long. . . . . .2. *ormocarpoides.*
      Pod not stipitate:
        Stamens 10 mm. or more long; calyx teeth very short; leaflets 5–9, often over 3 cm. long; endocarp spotted:
          Stem strongly 3–4-angled, the angles ± winged; racemes usually over 8 cm. long, including a peduncle usually over 2 cm. long; pod sometimes somewhat upturned, but not, or only slightly, thickened at the apex; hairs on leaf rhachis and lower surface of midrib ± spreading, those on upper surface of leaflets dense or sparse, parallel, appressed, medifixed, or, very rarely, absent . . . .
            3. *garckeana.*

          Stem not, or hardly triquetrous:
            Fruit not thickened at apex; peduncle usually over 2 cm. long; indumentum everywhere appressed, on the calyx so dense that it appears white, upper leaf surface sometimes glabrous . . . . .
              4. *binderi.*

           Fruit upturned and thickened at apex; racemes usually under 8 cm. including a peduncle usually under 2 cm. long; indumentum various; if dense enough for calyx to appear white then that on the leaf rhachis, etc., is usually spreading . . . . 5. *rhynchocarpa.*
           Upper leaf surface glabrous, or perhaps a few hairs on the veins:
             Indumentum on standard and calyx appressed . . . .
              α. var. *rhynchocarpa.*

            Indumentum on standard and calyx dense, ± spreading . . . . .
              β. var. *latipinna.*

           Upper leaf surface pubescent, the hairs spreading, irregularly arranged and largely basifixed . . . . . . γ. var. *uluguruensis.*
        Stamens less than 10 mm. long, or, if over 10 mm. (in Nos. 6 and 7), then leaflets more than 11:
          Hairs on fruit mostly dark brown or if (in No. 11) admixed with white, then very dense and ± spreading, fruit straight at all stages in its development, except sometimes for a slight upturn at the tip; endocarp spotted:
            Indumentum on outside of standard dense, very short, crisped, white; leaflets 13–19:
              Stamens *c.*10–12 mm.; brown hairs on calyx and midribs of leaflets spreading . . . . 6. *macrantha.*
              Stamens 7–9 mm.; brown hairs on calyx, pedicels, stems and midribs of leaflets ± appressed . . . . . . 7. *roseo-caerulea.*

Indumentum on outside of standard dense, appressed, golden brown-black:

Raceme almost sessile, usually shorter than subtending leaf; leaflets 3–9, rarely 11 or 13; calyx teeth much shorter than tube; hairs on leaves usually closely appressed, not dense enough to obscure leaf surface; plant usually drying brownish green:

Standard *c*.6–7 mm. long, persisting after carina and alae have fallen, sometimes even until the fruit dehisces; stamens 5–6 mm. long; leaflets usually 7–11 . . . . 8. *emarginella*.

Leaflets rounded-oblong or elliptic-obovate, not above twice as long as wide, rarely reaching 30 mm. in length . . . . α. var. *emarginella*.

Leaflets ± lanceolate, more than twice as long as wide, often over 30 mm. long . . . . β. var. *longefoliolata*.

Standard *c*.5 mm. long, falling off with the carina and alae soon after the flower opens, stamens *c*.4 mm. long, leaflets usually 3–5, suborbicular—rounded-oblong . . . . 9. *emarginelloides*.

Raceme ± pedunculate, usually longer than subtending leaf; calyx teeth about as long as tube:

Indumentum closely appressed, sparse, quite failing to obscure leaf surface; plant drying blackish; standard *c*.5 mm. long; stamens *c*.4 mm.; leaflets 7–9; racemes up to 15 cm. long, very slender . . . . 10. *macrophylla*.

Indumentum often ± spreading, often dense and obscuring leaf surface; plant often ± silvery; racemes not above 10 cm. long . . . . 11. *swaziensis*.

Stamens *c*.4–6 mm. long; peduncle *c*.1 mm. diam. leaflets 9–15 α. var. *swaziensis*.

Stamens *c*.3–4 mm. long; peduncle *c*.0·5–0·7 mm. diam.; leaflets 11–19 . . . . β. var. *perplexa*.

Hairs on fruit usually white; if some dark then sparse and closely appressed; fruit sometimes curved, especially when young, stamens not above 4 mm. long:

Petiole over 3 mm. long; leaves pinnate, or, if 3-foliolate, then pinnately so:

Fruit ± tetragonous; leaflets 3–9:

Indumentum of calyx and standard ± golden brown; old stems definitely woody with well marked lenticels, nearly black; leaflets 5–9; endocarp spotted . . . . . . 12. *subcorymbosa*.

Indumentum of calyx and standard white; lenticels if present obscure; endocarp not spotted:

Fruit very markedly tetragonous, ± straight, *c*.4 mm. from ridge to ridge, 3·5 mm. from suture to suture; leaflets 7–9 13. *lupatana*.

Fruit *c*.1·5–2 mm. across:

Pod arcuate (curved into a semicircle or beyond); leaflets 3, rarely 5 . . . . 30. *cliffordiana*.

Pod straight, sharply bent at the base or somewhat curved, but not arcuate:

Infructescence longer than subtending leaf; fruit (except
in No. 14 var. ε) hanging down alongside the axis
and only curving away from it towards the tip, if
at all:

Axis of inflorescence spinescent; fruit torulose, the
swellings in the plane of the sutures coinciding
with constrictions in the plane at right angles, and
vice versa; leaflets 3 . . . . 14. *spiniflora*.

Axis of inflorescence not spinescent, fruit not torulose,
or, if slightly so, then constrictions in both planes
coinciding . . . . 15. *subulata*.

Fruit as described in lines beginning "infructescence" above:

Leaflets 3:

Young shoots green, hairs appressed, leaflets
rounded or acute at apex. α. var. *subulata*

Young shoots silvery:

Hairs appressed; fruits long; leaflets emarginate . . . . β. var. *nubica*.

Hairs spreading; leaves smaller than usual;
leaflets not emarginate . . . .
γ. var. *microphylla*.

Leaflets 5–9, young shoots green . . . δ. var. *scabra*.

Fruit spreading stiffly outwards from a sharply bent
base; leaflets 3, not emarginate. ε. var. *maffeii*.

Infructescence shorter than subtending leaf; fruit spreading stiffly outwards from a sharply bent base;
leaflets 3 . . . . 16. *trita*.

Fruit not tetragonous; leaflets, apart from exceptional leaves, 5 or
more:

Pod torulose; seeds up to 5 or 6:

Stems densely argenteous (surface hidden by hairs); leaflets
more than 5 mm. wide; rhachis prolonged 3–4 mm.

beyond lateral leaflets:

Pods markedly torulose, the segments globular, dorsal and
ventral sutures narrow; leaflets 3–5 rarely 7, sericeous
above . . . . 17. *articulata*.

Pods less torulose, the segments oval-oblong, dorsal and
ventral sutures very wide; leaflets 5–9 rarely 11 . . . .
18. *coerulea*.

Leaflets glabrous above . . . . α. var. *coerulea*.

Leaflets with dense or scattered appressed hairs above
β. var. *occidentalis*.

Stems not densely argenteous (surface visible between hairs),
leaflets less than 5 mm. wide; rhachis prolonged less than
2 mm. beyond lateral leaflets:

Leaflets *c*.5–9 in number, rarely as much as 8 mm. long;
fruit *c*.2 mm. across at the bulges; infructescence
rarely over 15 mm. long . . . . 19. *bogdanii*.

Plant green; rhachis usually not prolonged beyond lateral
leaflets . . . . α. var. *bogdanii*.

Plant blackish; rhachis usually prolonged beyond lateral
leaflets . . . . β. var. *peteri.*

Leaflets usually more than 9 in number, usually over 8 mm.
long, if less than 9, then fruit *c.*3 mm. diam.;
infructescence 10–40 mm. long . . . . 20. *cryptantha.*

Swollen part of pod *c.*2 mm. across; fruit ± curved:

Leaflets up to 18 . . . . α. var. *cryptantha.*

Leaflets 18–27 . . . . β. var. *desmodioides.*

Swollen parts of pod *c.*3 mm. across, fruit straight, or
almost so . . . . γ. var. *occidentalis.*

Pod not torulose, or, if somewhat so, then seeds more than 6,
leaflets usually 11 or more:

Fruit short (under 10 mm.), thick (over 2 mm. wide), 1–3-
seeded; stems argenteous:

Indumentum closely appressed; endocarp obscurely spotted
21. *amorphoides.*

Indumentum of stems, leaf rhachis and fruits dense, crisped,
spreading . . . . 22. *cavallii.*

Fruit longer, more slender; seeds more than 3; endocarp
spotted:

Pod straight, *c.*6-seeded, less than 2 cm. long; indumentum
on flowers and inflorescence ± brown:

Leaflets 11–19, close together (*c.*5–9 mm. apart); infructescence short (rarely over 4 cm.) . . . . 23. *arrecta.*

Leaflets *c.*9–13, widely spaced (*c.*10–16 mm. apart);
infructescence very slender, up to 10 cm. long . . .
24. *longeracemosa.*

Pod ± curved; if short and *c.*6-seeded, then very definitely
curved, if almost straight, then more than 6-seeded and
over 2 cm. long; leaflets 9–19, ± closely spaced (under
10 mm. apart):

Fruit short, reddish brown, definitely curved, *c.*10–15 mm.
from end to end in a straight line, *c.*5–8-seeded;
hairs of inflorescence brownish . . . . 25. *suffruticosa.*

Fruit longer, greyish black, usually less curved or almost
straight, over 20 mm. from end to end in a straight
line, 8–12-seeded; hairs of inflorescence whitish
26. *tinctoria.*

Petiole under 3 mm. long; leaves ± digitately 3-foliolate (rarely
5-foliolate in No. 30) plant argenteous:

Young fruit straight:

Inflorescence not spinescent; leaflets up to 6 mm. long:

Calyx 1·5–2 mm. long; leaves fairly numerous. . . . 27. *ciferrii.*

Calyx 3 mm. long; leaves very few indeed . . . 28. *sparteola.*

Inflorescence spinescent:

Fruit flattened, not torulose; stems green; leaflets up to 6 mm.
long . . . . 29. *pungens.*

Fruit torulose; stems silvery; leaflets sometimes up to 15 mm.
long . . . . 14. *spiniflora.*

Young fruit arcuate; leaflets 8–18 mm. long . . . 30. *cliffordiana.*

92

Calyx lobes subulate, more than twice as long as tube; filaments 6·5–9 mm.
long; pod straight, more than 10-seeded; leaflets 3–9:

Stems angled, sometimes flattened, 2–3 mm. thick at base of inflorescences;
stipules up to 8 mm. long; stipellae well developed; leaflets up to 4 cm.
long; hairs at margin of leaflet larger and stouter than those of upper
surface, similar to those of lower surface; stamens *c*.9 mm. long;
inflorescences up to 40 cm. long . . . . *lasiantha.*

Stems cylindrical, *c*.1 mm. thick at base of inflorescences; stipules up to
2 mm. long; stipellae hardly visible; leaflets up to 2–2·5 cm. long;
hairs at leaflet margin similar to those of upper surface; stamens *c*.7 mm.
long; inflorescence up to 28 cm. long . . . . *fanshawei.*

Stipules ± foliaceous, 3 mm. or more wide; leaves digitately 3-foliolate on
petioles 1–4 cm. long; endocarp not spotted . . . . 31. *dimidiata.*

Stems creeping and rooting at the nodes; leaflets 11–17; stamens *c*.5 mm. long . . .
32. *repens.*

C.3.m.1. **I. podocarpa** *Bak.f. & Martin*, Bull J. Bot. Et. Brux. **8**, 154 (1923);
L.T.A. No. 223; F.C.B. **5**, 134.

"*I. macrophylla Schum. & Thonn.*" sensu Bak., F.T.A. **2**, 100 pro min. parte,
quoad pl. angolenses, sensu Hiern, Cat. Welw. Afr. Pl. **1**, 216, non Schum. &
Thonn.

OUBANGUI CHARI, Buar, *Mildbraed* 9794; B. CONGO, *Homblé* 1207 (BR), BM
type, *Quarré* 2444, etc.; TANGANYIKA, T4 Kahama distr., *Burtt* 5481; Tabora
distr., *Herring* 2012 EA; Mpanda distr., *Friend* 3158 EA; N. RHODESIA, N. Prov.,
Carson s.n. 16, *Burtt* 6123, *Mrs. Richards* 455, 660, 770, etc.; Central Prov.,
Mrs.Macaulay 915, *White* 2150; W. Prov., *Cruse* 479, *Eyles* 8123, *Marks* 23,
Milne-Redhead 3901, *White* 1826, 1868; ANGOLA, *Welwitsch* 2013 BM.

C.3.m.2. **I. ormocarpoides** *Bak.*, J. Linn. Soc. **22**, 464 (1887).

I. torulosa Bak., F.T.A. **2**, 91 (1871) non E. Mey (1836).

I. moniliformis Bak.f., J. Bot. **41**, 323 (1903); L.T.A. No. 149.

TANGANYIKA, T6, Ulugurus, *Miss E. M. Bruce* 847; PORT. E. AFR., Rovuma
riv., *Kirk* s.n., Lupata *Kirk* s.n. syntypes of *I. torulosa* Bak. and *I. moniliformis*,
Swynnerton 1439 BM, etc.; NYASALAND, *Buchanan* 1127; S. RHODESIA, *Wild* 4187;
N. RHODESIA, *Angus* 1023, *White* 2275, *St. Clair Thompson* 1341; MADAGASCAR,
Baron 4607, 4734, 4822 syntypes.

C.3.m.3. **I. garckeana** *Vatke*, Oest. Bot. Zeit. **29**, 221 (1879); Schweinf., Bull
Herb. Boiss. **4** app. **2**, 243 (1896); L.T.A. No. 192; F.C.B. **5**, 138.

I. tetragona Lebrun. & Taton, Expl. Parc. Nat. Kagera **1**, 72 (1948).

I. rhynchocarpa Welw. ex Bak. var. *quadrangularis* Berhaut, Fl. du Senegal 50
(1954) et Bull. Soc. Bot. Fr. **101**, 375 (1955).

SENEGAL, Ouassadou, *Berhaut* 1219 P holotype of *I. rhynchocarpa* var. *quad-
rangularis*; B. CONGO, East, *Lebrun* 9584 BR holo. K isotype of *I. tetragona*, etc.;
A.-E. SUDAN, *Schweinfurth* 3748, 3785; ERITREA, *Pappi* 351; ETHIOPIA, South,
Gillett 14491, *Bally* 9248; UGANDA, U1–4; KENYA, K3 Cherangani, K4 (*Hildebrandt*
2795 holotype), K5; TANGANYIKA, T1–4, T6; PORT. E. AFR.; S. RHODESIA;
N. RHODESIA.

This species is close to *I. rhynchocarpa*, and intermediates occur where their
areas overlap, although rarely. As most specimens do not show mature fruit the
curious terminal thickening of the fruit of *I. rhynchocarpa* is not certainly a good
character. Out of some 50 specimens of *I. garckeana* examined (sorted on the basis
of their winged stems) 2 only (*Chancellor* 249 from West Nile, U1 and *Milne*

93

Redhead & Taylor 7371 from Morogoro, T6) lacked the usual appressed medifixed hairs on the upper leaf surface. This type of indumentum was not seen at all in the somewhat larger number of specimens of *I. rhynchocarpa* which were examined. Taking this correlation of character into account, it seems that the specific distinction between *garckeana* and *rhynchocarpa* should be maintained. *I. garckeana* in some of its forms also comes very close to *I. binderi* and further careful collecting is necessary to establish the relationship between these two taxa.

C.3.m.4. **I. binderi** *Kotschy*, Sitz. Ak. Wien. Math. Nat. **51** abt. **2,** 364 t. 6b (1865); L.T.A. No. 193.

Oubangui Chari, *Chevalier* 6603; A.-E. Sudan, *Binder* s.n. W. holotype, *Schweinfurth* 1534, 2525, *Myers* 8469, *Andrews* 538, *Jackson* 2257; Kenya, K1 *Gillett* 12903, 13488, *Bally* 9115.

C.3.m.5. **I. rhychocarpa** *Welw. ex Bak.*, F.T.A. **2,** 92 (1871); L.T.A. No. 188; F.C.B. **5,** 136.

α. var. **rhynchocarpa.**

I. lonchocarpifolia Bak., Kew Bull. **1897,** 256; L.T.A. No. 194.

I. flavovirens R. E. Fries, Wiss. Ergeb. Schwed. Rhod. Kongo Exp. **1,** 79 (1914); L.T.A. No. 189.

I. lignosa De Wild., Pl. Beq. **3,** 323 (1925); L.T.A. p. 879.

I. gyrocarpa Bak.f., J. Bot. **73,** 78 (1935).

Oubangui-Chari, *Mildbraed* 9332, 9800, *Tisserant* 3345; B. Congo, *Claessens* 1103 BR holotype of *I. lignosa*, etc.; Tanganyika, T1, T3–4, T5 *Burtt* 1734 BM, 3452 BM syntypes of *I. gyrocarpa*, T6–8; Port. E. Afr.; Nyasaland, *Whyte* s.n. holotype of *I. lonchocarpifolia*; S. Rhodesia; N. Rhodesia, *R. E. Fries* 1211 UPS holotype of *I. flavovirens*; Angola, *Welwitsch* 2015 type.

A variable plant. The type has a sparse appressed indumentum; other specimens show a denser spreading indumentum especially on the stems, leaf rhachis and veins. The size and shape of the leaflets also varies greatly. These characters however, vary so continuously that it does not seem worthwhile to distinguish further infra specific taxa.

β. var. **latipinna** (*Johnson*) *Gillett* stat. nov.

I. latipinna Johnson, Contr. Gray Herb. n.s. **73,** 35 (1924); L.T.A. No. 190.

N. Rhodesia, copper belt, *Allen* 383 (compared with type by Mr. Milne-Redhead), *Brenan & Greenway* 7967 K, EA, *Cruse* 26, 451, *Fanshawe* 520; Angola (*Curtis* 360 GH holotype).

Indumentum of stems, lower leaf surface, leaf rhachis, inflorescence and calyx dense, white, ± spreading. That on the standard dense, golden brown, ± spreading. Leaflets longer than usual. Some forms of var. *rhynchocarpa* approach this plant, but, as here delimited the indumentum of calyx and standard is in that variety always appressed. *I. nambalensis* Harms in Warburg & Baum Kunene-Zambesi Exp. 258 (1903); L.T.A. No. 191; type (*Baum* 254 B†), may belong to this variety but the description is not full enough to be sure. The type of var. *latipinna* shows no fruit This is seen in *Cruse* 451; it is upturned and swollen at the tip as in var. *rhynchocarpa* but is tomentose.

γ. var. **uluguruensis** *Gillett* var. nov.

A varietatibus aliis differt foliolis superne pilis patentibus saepius basifixis indutis. Indumentum etiam quam illud var. *rhynchocarpae* ubique densius.

Tanganyika, T3 Handeni distr., 32 km. S. of Handeni, 600 m., *Brachystegia* woodland, shrub 1 m. tall, standard pale green with dark hairs outside, keel pale green, wing petals white or faintly greenish, 5.4.1953, *Drummond & Hemsley* 2032:

T6 Morogoro distr., 900 m., softly woody, 1·7 m. tall, flowers green with brown hairs, Swahili name Msango Sango, 16.10.1932, *Wallace* 217; in young fruit 6.1.1933, *Wallace* 217A holotype; abundant, 21.10.1934, *Miss E. M. Bruce* 23; 5.12.1934, *Bruce* 289; 1400 m., 1.11.1914, *Peter* K82; in high grass—low tree savannah in young fruit, May 1938, *Wigg* 1215 EA. Morogoro, above Agricultural nurseries, 600 m., in *Brachystegia* woodland on steep slopes, weak shrub up to 2 m. tall, standard and keel greenish white with brown hairs outside, wings white sometimes tinged with pink, petals falling at 9 a.m., 24.11.1955, *Milne-Redhead & Taylor* 7372.

The indumentum on the upper leaf surface is quite different from that in *I. garckeana*. The fruits are markedly thickened at the apex, and pubescent. The stem is free from any trace of a wing.

C.3.m.6. **I. macrantha** *Harms*, Ann. Ist Bot. Rom. **7**, 91 (1898); L.T.A. No. 227.

Ethiopia, nr. Agheremariam, *c.*5°38′N, 38°14′E, 16.9.1893, *Riva* 1313 FI holotype.

Harms described the ovary as glabrous, but this is a mistake; it has a dense appressed dark brown indumentum as in the next species.

C.3.m.7. **I. roseo-caerulea** *Bak.f.*, L.T.A. 158, No. 224 (1926); F.C.B. **5**, 137.

B. Congo, East, fide Cronquist; Ethiopia, Djem Djem, Wouramboulchi, *Owen Cooper* s.n.; Uganda, U3 Elgon *Dümmer* 3593 holotype, *Tothill* 2398, 2682, etc.; Tanganyika, T2 *St. Clair Thompson* 439; T7 *Stolz* 1141, *Ward* U16, *Carmichael* 170; T8 *Zerny* 517 W.

Doubtfully distinct from *I. macrantha*.

C.3.m.8. **I. emarginella** *Steud ex A. Rich.*, Tent fl. Ab. **1**, 184 (1847); L.T.A. No. 221; F.C.B. **5**, 135.

α. var. **emarginella.**

I. coluteifolia Jaub. & Spach, Ill. Pl Or. t. 487–8 (1856).

I. cameroni Bak.f., J. Linn. Soc. **15**, 93 (1876); L.T.A. No. 222.

N. Nigeria; Oubangui-Chari; B. Congo; A.-E. Sudan; Ethiopia, *Schimper* (1840) 1723 isotype of *I. emarginella* and *I. coluteifolia*; Uganda, U1–4, Kenya, K3, K5; Tanganyika, T1, T4 *Cameron* 67* holotype of *I. cameroni*, T5–8; Port. E. Afr.; Nyasaland; S. Rhodesia; N. Rhodesia; Angola.

β. var. **longefoliolata** *Gillett* var. nov.

A var. *emarginella* foliolis angustioribus, longioribus paucioribus basin attenuatis differt.

N. Rhodesia, Abercorn distr., *c.*1500 m., Scott's farm, red sandstone and loam soil, woodland, ± woody, 25 cm. tall. standard dark brown, wings light red, keel pale, 11.1.1955, *Mrs. Richards* 4007. Kasulo, in open woodland, 75 cm. tall. 11.1.1955, *Mrs. Richards* 4029. Chilongowelo, bush above Victoria falls, *c.*1400 m., in rather thick undergrowth in gorge, 28.1.1955 (young fruits collected later) *Mrs. Richards* 4230B. holotype. Bush by Tasker's deviation, in open bush among grass on slope, *c.*130 cm. tall, 23.1.1955, *Mrs. Richards* 4233. May 1955 in fruit, *Mrs. Richards* 4233/B.

The leaflets which may be up to 8 cm. long and 4 times as long as wide make this plant so strikingly different in appearance from the usual *I. emarginella* that a varietal name seems called for. The leaflets are often 3 only and seem not to be more than 9. *I. emarginella* is in general a variable species in Tanganyika, and in N.

Rhodesia. The hairs on the calyx may be nearly white, as in *Cameron 67** instead of the usual brown, but are always dense.

C.3.m.9. **I. emarginelloides** *Gillett* sp. nov.

I. emarginellae affinis; ab ea ramis floriferis tenuioribus, foliolis paucioribus, racemis brevioribus, floribus minoribus, indumento calycis sparsiore, vexillo mox caduco differt.

Fruticulus *c.*1 m. altus, cortice ramorum vetustiorum griseo, ramis novellis floriferis tenuibus (0·5–1·0 mm. diametro) flexuosis, viridibus, pilis saepius fuscis, medifixis, crispis [vel appressis] indutis. *Folia* 1–2 [–3]- jugata; stipulae anguste deltoideae vel sublineares, pilis medifixis fuscis et albidis indutae, ad 3 mm. longae; rhachis superne canaliculata, basin pulvino atroviride munita, pilis saepius fuscis medifixis induta, ultra foliola lateralia opposita ad 13 mm. prolongata, in tota 10–35 [–60] mm. longa, petiolo 5–15 mm. longo incluso; stipellae filiformes virides vel fuscae, 1 mm. longae; petioluli atrati, 1–1·5 mm. longi; foliola elliptica, emarginata, apiculata, utrinque pilis medifixis haud dense induta, pilis inferne ad nervos brunneis, alibi albidis, 10–13 [–30–60] mm. longa, 7–11 [–24–40] mm. lata. *Racemus* ad 20-florus, fere sessilis, 10–20 mm. longus, pilis saepius albidis vestitus, bracteis *c.*1 mm. longis, mox deciduis, pedicellis *c.*1 mm. longis, post anthesin ± reflexis. *Calyx* pilibus albidis haud dense indutus, tubo inaequale superne 1 mm. longo, inferne 1·5 mm. longo, dentibus subulatis inaequalibus, 2 superioribus 0·6 mm. longis, 3 inferioribus 1 mm. longis, sinibus inaequalibus, superiore 2 mm. lato, lateralibus 0·9 mm. latis, inferioribus 0·6 mm. latis. *Vexillum* externe pilis brunneis appressis dense obsitum, late ellipticum, acuminatum, 4 mm. latum, 5·6 mm. longum, cum alis carinaque mox deciduum. *Alae* ad margines pilosae, ± oblongae, 1–1·5 mm. latae, 4·8 mm. longae, unguiculo 0·5 mm. longo incluso. *Carina* extus pubescens, subrostrata, *c.*1 mm. calcarata, 5·5 mm. longa. *Stamina* omnia fertilia, filamentis 3–4 mm. longis, antheris ellipticis, basin pilosis, c.0·8 mm. longis, appendice apicale 0·2 mm. longo incluso. *Ovarium* 3 mm. longum, in stylum 1·5 mm. longum, 1 mm. infra stigma globosum 0·05 mm. diametro rectangulare curvatum, sensim attenuatum, *c.*5-ovulatum. *Legumina* in quoque racemo pauca (1–5), iuventute nigra, demum brunnea, appresse breviter pilosa, teretia, acuta, recta, apice paulo curvato excepta, 24–28 mm. longa, 2 mm. lata et crassa, endocarpio valide atrorubre maculata. [*Semina* *c.*6 brunnea, subpunctata, *c.*2 mm. longa, 1·5 mm. lata, 1 mm. crassa, inter se *c.*1·5 mm. distantia.]

Tanganyika, T4 Tabora, Simbo reserve, 1200 m., June 1938 (in fruit), *Wigg* 1174 EA; N. Rhodesia, N. Prov., Abercorn district between Mbete and Izi falls, 900 m., rocky gorge clothed with *Burttia, Brachystegia holstii*, etc., locally common, *c.*1 m. tall, 8.5.1936, (in fruit), *Burtt* 5991/A (5991 with which this was mixed is *I. subcorymbosa*); path from Posa village to Mpulungu road, 720 m., woodland among grass, flowers insignificant, pale pink, stem woody, 16.12.1954, *Mrs. Richards* 3654; Chilongowelo, path to dogs' graves, 1500 m., in bush under trees, soil red, flowers pink, falls as soon as it opens, 22.12.1954, *Mrs. Richards* 3737. Path to Inono source, 1500 m., shrubby, 30 cm. tall, in fruit 9.5.1955, *Mrs. Richards* 5614. Mpika, Plateau woodland, subshrub 65 cm. tall, in fruit 28.1.1955, *Fanshawe* 1879, Copper belt, Ndola east forest reserve, small wiry shrub 1 m. tall, Chi Wemba name Kaseketi, 4.1.1952, *F. White* 1867; Mwekere, Kitwe, in plateau woodland, subshrub 70–100 cm. tall, flowers greenish white, 5.12.1953, *Fanshawe* 532 holotype; 12.5.1955, in fruit, *Fanshawe* 2277. Mwinilunga district 1·5 km. S. of Matonchi farm, in *Brachystegia* woodland on sand, shrubby herb 1 m. tall, 31.12.1937, *Milne-Redhead* 3904.

This species seems closely related to *I. emarginella* and it is noteworthy that in the area occupied by both species, and not elsewhere, characters such as 1–3-foliolate leaves and white hairs on the calyx occur occasionally in *emarginella*, which suggests that introgressive hybridisation may occur. The following table sets out the differences between the two species.

I. emarginella	*I. emarginelloides*
1. Leaves 1–6-, usually 3–5-jugate.	Leaves 1–3-, usually 1–2-, jugate.
2. Floriferous stems usually more than 1·5 mm. wide, often somewhat ridged, often erect from a perennial rootstock. Buds of next year's shoots rarely apparent.	Floriferous stems usually less than 1 mm. wide, not or hardly ridged, usually short shoots from an older stem. The dark buds of next year's shoots obvious in fruiting specimens.
3. Inflorescence often producing 50 or more flowers and up to 10 fruits, usually over 2 (up to 7) cm. long; its axis strongly ridged and much thicker than a fruiting pedicel; the latter strongly reflexed.	Inflorescence producing up to 20 flowers and rarely more than 3–5 fruits, rarely over 2 cm. long; its axis hardly ridged and hardly thicker than a fruiting pedicel, the latter less strongly reflexed, fruits occasionally ascending.
4. Indumentum of calyx dense (epidermis not visible between hairs) usually brown, rarely white.	Indumentum of calyx not dense (epidermis readily visible between hairs) white.
5. Standard persistent long after wings and keel have fallen.	Standard soon falling along with wings and keel.
6. Stamens *c.*5 mm. long.	Stamens *c.*4 mm. long.

I. emarginelloides is itself rather variable especially in the density of the indumentum, whether it is appressed or crisped, and leaf size. *Mrs. Richards* 3654 is a weaker less erect plant than usual, *Burtt* 5991/A has much longer leaflets than usual.

C.3.m.10. **I. macrophylla** *Schum.*, Beskr. Guin. Pl. 372 (1829); L.T.A. No. 225; non sensu Hiern, Cat. Welw. Afr. Pl. **1**, 216.

"*I. forsan binderi Kotschy*" sensu Chevalier, Expl. Bot. Afr. Occ. Fr. 172 non Kotschy.

Senegal, *Chevalier* 3238 P; Port Guinea; Fr. Guinea; Sierra Leone; Liberia; Ivory Coast (fide Tisserant); Gold Coast (*Thonning* s.n. C. holotype); Togo; Dahomey, *Le Testu* 213 P; Nigeria; Brit. Cameroons.

The Angolan plant identified as *I. macrophylla* in F.T.A. **2**, 100 (*Welwitsch* 2013) is *I. podocarpa*. Baker's statement, repeated by E. G. Baker in L.T.A. that the fruits of *I. macrophylla* are glabrous is erroneous.

C.3.m.11. **I. swaziensis** *Bolus*, Tr. S. Afr. Phil. Soc. **16,** 381 (1906).

α. var. **swaziensis.**

I. mearnsii Standley, Smithson. Misc. Coll. **68**: 5, 5 (1917); L.T.A. No. 231.

I. oliveri Schweinf. ex Engl., Hochgebirgsflora Afrikas, 259 (1892); L.T.A. 230 (nomen).

I. oliveri Schweinf. ex Harms, Not. Bot. Gart. Berl. **10**, 81 (1927).

A.-E. Sudan, Lorienaton, *Dale* 323; Kenya, K3 Kinangop, *Battiscombe* 951; K4 *Mearns* 208, 958, 1004 US holotype of *I. mearnsii*, nr. Nyeri (*Von Hoehnel* 75 B† holotype of *I. oliveri*), R.E. & T.C.E., *Fries* 417 K, UPS, neotype of *I. oliveri* etc.; K5 *Napier* 2949, etc.; K6 *Bally* 4145, etc.; Tanganyika, T1 *Conrads* 5758; T2 *Tanner* 1783; Transvaal, *Wilms* 315 BM, 327, *Schlechter* 3908 syntypes, *Pott* 5320, *Liebenberg* 3320; Swaziland (*H. Bolus* 11836 BOL syntype).

β. var. **perplexa** (*N. E. Brown*) *Gillett* stat. nov.

I. perplexa N. E. Brown, Kew Bull. **1925**, 154.

Tanganyika, T2 *Johnston* s.n., *Peter* 43359, etc.; T3 *Greenway* 2082, etc.; T5 *Hornby* 446; T6 *E. M. Bruce* 1112, *Burtt* 4417, etc.; T7 *Stolz* 735; T8 *Zerny* 388 W; Port. E. Afr., Zambezia Nhamarroe, *Torre* 5449 LISC, Manica & Sofala,

Chimoio *Garcia in Mendonça* 799 K, LISC; S. RHODESIA, Lomagundi distr., *Jack* 257, Umtali, *Teague* 610, *Eyles* 7087; TRANSVAAL, *Junod* s.n. holotype, *Hutchinson* 2185, *Smuts* 302, etc.

Varieties α and β intergrade into one another much more in East Africa than in the Transvaal. Harms l.c. mentioned var. β as a form of *I. oliveri* without naming it. As Harms remarks, this species is often difficult to distinguish from *I. emarginella*. It appears that, in East Africa at any rate the areas occupied by the two species hardly overlap. Standley correctly describes *I. mearnsii* as having dark hairs on the fruits, he is wrong however in contrasting it in this respect with *I. macrophylla* which, in fact has similar hairs on its fruits. E. G. Baker wrongly states that *I. mearnsii* has glabrous fruits. This species sometimes has fruits ± erect or spreading: it is however, far too close to *I. emarginella* to be placed outside the present subsection.

C.3.m.12. **I. subcorymbosa** *Bak.*, F.T.A. **2**, 91 (1871); L.T.A. No. 186, including var. *eylesii* Bak.f., L.T.A. 149 (1926).

"*I. sp. near subcorymbosa Bak.*" sensu Brenan Check List T.T. trees and shrubs 429 (1949).

TANGANYIKA, T4 *Humbert* 7184, *Burtt* 3451, 3583; T5 *Anderson* 308, *Burtt* 3610, 4575, etc.; PORT. E. AFR., Nyassa Prov. *Pedro & Pedrogao* 3379 EA; NYASALAND, *Buchanan* 403; S. RHODESIA, *Seymour Hall* in SRGH 33027, *Teague* 605, *Eyles* 1958 type of var. *eylesii*; N. RHODESIA, N. Prov., *Burtt* 5991 K, EA, *Mrs. Richards* 1063; West Prov. *Duff* 89/33; ANGOLA, *Welwitsch* 2053 type.

NOTE: The type has leaves with 5–7 leaflets; specimens from Tanganyika and S. Rhodesia occasionally show leaves with 9 leaflets; specimens from N. Rhodesia have leaves with 9–15 leaflets which tend to be larger than those from the other areas.

C.3.m.13. **I. lupatana** *Bak.f.*, L.T.A. 154, No. 210 (1926).

I. goniocarpa Bak.f., J. Bot. **71**, 341 (1933).

I. commiphoroides Chiov. in Cufodontis Miss. Biol. Boran 69 t. 11 (1939).

S. ETHIOPIA, Arero, *Cufodontis* 340 FI holotype of *I. commiphoroides*; KENYA, K1 *Gillett* 12935, 13063, 13491; K4 *Graham* 2100, 24400, etc.; K6 *Napier* 2388; TANGANYIKA, T1 *Tanner* 1490; T2 *Peter* 44068, etc.; T3 *Holst* 2690A, etc.; T4 *Wigg* 1172 EA; T5 *Burtt* 3572 BM holo K isotype of *I. goniocarpa*, etc.; T6 *Haarer* 1952; PORT. E. AFR., Lupata, *Kirk* s.n. holotype K, LISC, *Torre* 1964, 4295 LISC; NYASALAND, *White* 2534; S. RHODESIA, *Wild* 2757, *Chase* 4706, *R. M. Hornby* 3269, *McGregor* 1722/51; N. RHODESIA, *Robinson* 658; TRANSVAAL, Kruger Nat. Park, *Codd* 4909, 5259, *H. P. v.d. Schyff* 3606.

The type of *I. lupatana* has relatively large leaflets (up to 21 mm. long) while those of *I. goniocarpa* and *I. commiphoroides* are much smaller (up to 8 and 4 mm. respectively). However all gradations exist and the differences are very probably a matter merely of the amount of water available.

C.3.m.14. **I. spiniflora** *Hochst. & Steud. ex Boiss.*, Fl. Or. **2**, 190 (1872) in obs.

I. spiniflora Hochst. & Steud in Schimp. Pl. Arab. No. 945 (nomen) et Steud., Nomencl. ed. 2,808 (1840) (nomen.).

"*I. spinosa Forsk.*" sensu Bak., F.T.A. **2**, 77 (1871) pro min. parte, non Forsk.

I. spinosa Forsk var. *spiniflora* Schweinf., Bull. Herb. Boiss. **4**, app. **2**, 237 (1896) p.p. (nomen); Chiov. in Pirotta Fl. Eritrea, Ann. 1st Bot. Rom. **8**, (1903) 91 cum. descr.; Bak.f., J. Bot. **41**, 236 (1903); L.T.A. p. 113.

HEJAZ, near Jeddah, *Schimper* (1837) 945 isotype, *Fischer* 76, *Zohrab* 59, *Trott* 1543; ADEN, *Beevor* 54; A.-E. SUDAN, Red Sea Coast *c.*22°N., *Schweinfurth* s.n. K,

BM, *Shabetai* F 1053; *c*.21°N. *Bent* s.n. p.p.; *c*.18N. *Bally* 6962; ERITREA, *Popov* 1404; BRIT. SOMALILAND, Zeila, *McKinnon* S.50.

Chiovenda l.c. remarks that "the characters which distinguish *I. spiniflora* are sufficiently important and constant to justify its recognition, at least as a variety." The fact that some 18 characters, as summarised below, separate *I. spiniflora* from *I. spinosa* and that, among all the material now available, only two possibly intermediate specimens have been seen, seems quite enough to justify Hochstetter and Steudel's original view that it is a distinct species. It probably is not even particularly closely related to *I. spinosa*, the apparent resemblances being due to convergence in a desert environment. It seems much more closely related to *I. subulata* and *I. trita*.

I. spiniflora	*I. spinosa*
1. *Petiole* 2–6 mm., leaves always 3-foliolate.	*Petiole* 2–4 mm., leaves occasionally 5–7-foliolate.
2. *Mature inflorescence spines* green;	*Mature inflorescence spines* golden brown;
3. up to 5 cm. long in luxuriant plants;	rarely exceeding 2 cm.;
4. rather weak and not very prickly at the tip.	strong and prickly at the tip.
5. *Flowers* up to 20 per inflorescence;	*Flowers* 2–4 per inflorescence;
6. borne throughout its length.	usually borne above its middle.
7. *Calyx* lobes longer than tube.	*Calyx* lobes about as long as tube.
8. *Standard c*.5×2·5 mm.,	*Standard c*.5×4 mm.,
9. persistent,	not persistent,
10. its indumentum closely appressed.	its indumentum rather spreading.
11. *Pods* densely appressed-argenteous,	*Pods* sparsely appressed-strigose,
12. peculiarly torulose (as described in key),	not torulose,
13. strongly reflexed,	suberect, spreading or sub-reflexed,
14. almost indehiscent;	freely dehiscent,
15. endocarp not spotted.	endocarp spotted.
16. *Seeds c*.0·6 mm. in plane of suture, *c*.1·1 mm. in plane at right angles,	*Seeds c*.1·5 mm. in plane of suture, *c*.1·0 mm. in plane at right angles to this,
17. hardly pitted;	pitted;
18. streaked with black.	not streaked with black.

Three rather imperfect specimens from British Somaliland, *Gillett* 4532, *Collenette* 82, *Bally* 7243 may represent a form, or forms of this species. They also show some affinity with *I. sparteola* Chiov. and *I. ciferrii* Chiov. Further collections of the suffruticose, argenteous, ± spinescent *Indigoferas* from this region are needed. Klotzsch in Peters Moss. Bot. corrigenda and Schweinfurth in Fl. Aethiop. 13 state that *Tephrosia crotalarioides* Klotzsch l.c. 45 is the same as *I. spiniflora*. However the description does not agree, e.g. petals said to be glabrous in *T. crotalarioides*; and, until now, true *I. spiniflora* has not been seen within thousands of miles of Sena in Port. E. Afr. where *T. crotalarioides* was collected. Schwarz in Mitt. Inst. Allgem. Bot. Hamburg **10**, 102 (1939) states that var. *spiniflora* sensu Schweinfurth is partly *I. tritoides* Bak. and no reliance can be placed on Schweinfurth's identification of *Schweinfurth* (1891) 1282 from 1500 m. in Eritrea, with "leaflets up to 13 mm. long" and *Schimper* and *Ehrenburg* specimens cited in Flor. Aethiop. 13 (1867).

C.3.m.15. **I. subulata** *Vahl ex Poir.* in Lam. Encyc. Suppl. **3,** 150 (1813); L.T.A. No. 172; Meikle, Kew Bull. **1950,** 352; F.C.B. **5,** 161.

α. var. **subulata.**

I. thonningii Schum. & Thonn., Beskr. Guin. Pl. 366 (1829).

I. tetragonoloba E. Mey., Comm. Pl. Afr. Austr. 106 (1836).

"*I. trita L.f.*" sensu Bak., F.T.A. **2,** 86 (1871) non L.f.

I. dimorphophylla Schinz, Abh. Bot. Ver. Brand. **30,** 164 (1888); L.T.A. No. 171.

I. carinata De Wild., Bull. J. Bot. Et Brux. **8,** 147 (1923).

I. subincana N. E. Brown, Kew Bull. **1925,** 155.

Senegal; Sierra Leone; Liberia; Gold Coast, *Thonning* s.n. (P holotype) (C isotype) Photo K, *Thonning* s.n. (C holotype of *I. thonningii*) photo K; Togoland; Nigeria; Fr. Eq. Afr.; B. Congo, *Bequaert 5383* BR holotype of *I. carinata*; A.-E. Sudan; Ethiopia; Uganda, U2; Kenya, K4–7; Tanganyika, T1, T3, T5, T6; Zanzibar; Pemba; Port. E. Afr.; S. Rhodesia; Angola; S.W. Africa, *Schinz 246* isotype of *I. dimorphophylla*; Transvaal, *Rogers 22560* holotype of *I. subincana*; Natal (*Drège* s.n. type of *I. tetragonoloba*).

I. dimorphophylla and *I. subincana* differ somewhat in their rather more silvery indumentum, but numerous intermediates occur.

β. var. **nubica** *Gillett* var. nov.

A var. *subulata* indumento ramulorum dense argenteo, foliolisque apice truncatis vel emarginatis differt.

A.-E. Sudan, coastal prov., 22°N, Ssoturba mts. *Schweinfurth* (1865) 1812 K holo., BM isotype, J. Elba, high up on mt., a small bush, *Shabetai* F 1893, *B. K. Cooke* 138; 21°N *Bent* s.n.

Mrs. Lort Phillips s.n. (1897) from Brit. Somaliland may also belong to this variety.

γ. var. **microphylla** *Chiov.*, Flora Somala **1,** 138 (1929).

Somalia, *Puccioni & Stefanini 319* (366) FI holotype.

δ. var. **scabra** (*Roth*) *Meikle*, Kew Bull. **1950,** 352 (1951).

I. scabra Roth, Nov. Pl. Sp. 359 (1821).

I. mucronata Spreng. ex DC., Prodr. **2,** 227 (1825), non Lam.

? *I. jamaicensis* Spreng, Syst. Veg. **3,** 277 (1826)?

I. flaccida Koenig. ex Roxb., Fl. Ind. **3,** 375 (1832).

I. quartiniana A. Rich., Tent. Fl. Ab. **1,** 183 (1847); L.T.A. No. 166 p.p.

I. coronilloides Jaub. & Spach, Ill. Pl. Or. t.t. 485–6 (1856) non Mart.

I. spachii Bak., F.T.A. **2,** 92 (1871); L.T.A. No. 165.

? *I. umbraticola* Vatke, Oest. Bot. Zeit. **28,** 202 (1878); L.T.A. No. 197.

I. retroflexa Baill., Bull. Soc. Linn. Par. **1,** 399 (1883); L.T.A. No. 168; *Edwards & Bogdan*, Impt. Grassland Pl. Kenya fig. 31; 3.

I. rutschuruensis De Wild., Bull. J. Bot. Et Brux. **8,** 155 (1923); L.T.A. No. 167.

Peninsular India, probably Madras Prov. Wallich coll. 5475 ex herb Heyne isotype of *I. scabra*, Icon Roxburghiana 384 K type of *I. flaccida*; Nigeria; B. Congo, *Bequaert 6279* BR holotype of *I. rutschuruensis*; Ethiopia, 1852–3 (*Schimper* s.n. P syntypes of *I. coronilloides* and *I. spachii*) (*Quartin Dillon* P syntype of *I. quartiniana*); Br. Somaliland, *Hildebrandt 1386* isotype of *I. umbraticola*; Uganda, 2–4; Kenya, K4, K6–7; Tanganyika, T2–3, T6–7; Zanzibar; Port E. Afr.; Nyasaland; S. Rhodesia; Madagascar, *Grandidier* (P holotype of *I. retroflexa*) K photo; Transvaal; Central America; West Indies, Jamaica (*Bertero* s.n. type of *I. jamaicensis*) (*Bertero* s.n. type of *I. mucronata* Spreng. ex DC. non Lam.)

Note: *Hildebrandt 1386* is the only specimen seen from Brit. Somaliland; the Kew sheet shows young flowers and buds only. Vatke describes the fruits as having only 2 seeds, perhaps as a result of abortion. If this is found to be normal in the Somaliland plant and not merely a casual occurrence it may be necessary to treat *I. umbraticola* as a distinct taxon.

ε. var. **maffeii** (*Chiov.*) *Gillett* stat. nov.

I. quartiniana A. Rich., Tent. Fl. Ab. **1**, 183 (1847) p.p.; L.T.A. No. 166 p.p.

"*I. marginulata Graham*" sensu Bak. in Hook.f. Fl. Brit. Ind. **2**, 97 (1876) p.p. non Graham ex Wight and Arn.

I. spachii Bak. var. *trifoliolata* Schwnf., Bull. Herb. Boiss. **4**, 240 (1896) p.p.

I. maffeii Chiov., Att. Ist Bot. Pav. Ser. **4. 7**, 128, tab. 5 (1936).

Foliola 3. A varietatibus aliis *I. subulatae* fructibus rectis, basin a rhachide valde abrupteque deflexis, sub angulo 70°–90° divagantibus, differt. Quoad fructo *I. tritae* similis, sed racemis multo laxioribus, quam folium longioribus, differt.

INDIA, Punjab, *Drummond* 23658, Konkan *Stocks* s.n.; A.-E. SUDAN, *Babiker Beshir* 142, *Lea* 98, *Andrews* 98, 121, 132; ERITREA, *Schweinfurth & Riva* 648 isosyntype of *I. spachii* var. *trifoliolata*, *Pappi* 3944 K, BM, 4056 BM; ETHIOPIA, Chiré, *Quartin Dillon* s.n. syntype of *I. quartiniana*; SOMALIA, Genale (*Pollacci & Maffei* 124 FI holotype); KENYA, K1 *Gillett* 13092 K, EA; TANGANYIKA, T1, *Rounce* 48 K, EA; PORT. E. AFR., Sul do Save, *J. Pedrogao* 279.

This is the plant to which Meikle 1.c. draws attention as a connecting link between *I. subulata* and *I. trita*. It might equally well be assigned to either species. As in var. *subulata* and var. *scabra* there is much diversity within the variety in leaf size, etc. It appears that both *I. quartiniana* A. Rich, and *I. spachii* var. *trifoliolata* Schweinfurth include more than one of the varieties of *I. subulata* which are here recognised, their authors having paid no attention to the peculiar fruit of var. *maffei*.

I. strigulosa Bak.f., L.T.A. 143, No. 162 (1926).

PORTUGUESE CONGO, Alto dos Cruzes, *Gossweiler* 5882 BM holotype, is probably but a form of *I. subulata* var. *subulata*. It differs in the denser, stiffer indumentum. In the absence of fruits the status of this plant must remain doubtful.

I. subulata is often a weed of cultivation and its range has probably been extended on this account.

C.3.m.16. **I. trita** *L.f.*, Suppl. 335 (1781); L.T.A. No. 164

INDIA; CEYLON; BURMA; TIMOR; AUSTRALIA (see note).

NOTE: Some of the Australian sheets are atypical, several have 3–5 leaflets instead of consistently 3 as in India and another has a very long inflorescence. This species has often been recorded from Tropical Africa but the plants in question belong to one form or another of *I. subulata*. To some extent the distinction is artificial as, if more weight were attached to the shape of the pod and less to the length of the inflorescence, var. *maffeii* would have to be associated with this species instead of with *I. subulata*.

C.3.m.17. **I. articulata** *Gouan*, Illus. 49 (1773); L.T.A. No. 207 p.p.; Andrews, F. Pl. A.-E. Sudan **2**, 216 p.p.

"*I. argentea* L." sensu Bak., F.T.A. **2**, 97 p.p.? L., Mant. **2**, 273 (1771), excl. syn., non. *I. argentea* Burm. f. (1768).

"*I. tinctoria*" sensu Forsk., Fl. Aeg. Arab. 138 (1775) p.p. non L. (1753).

I. glauca Lam., Encyc. **3**, 246 (1789).

Plant of unknown origin cultd. at Paris (P), K photo, holotype of *I. glauca*.

EGYPT, *unknown coll.* ex herb. Gouan holotype, *Boué* s.n. *Delile* s.n. BM, *Aucher-Eloy* 1017, *Bromfield* s.n., *Schimper & Wiest* (1835) 57 K, BM, *Schweinfurth* (1864) 1486 BM; JORDAN and adjacent areas, *Post* 11, *Boué* 282, *Hart* s.n. K, BM, *Lowne* s.n., *Meyers and Dinsmore* 5851; SAUDI ARABIA, *Ehrenburg* s.n., *Philby* 103 BM, S.E. Rub el Kali *Thesiger* s.n. BM; YEMEN, Ta 'iz, *Scott & Britton* 346 BM; OMAN, *Aucher-Eloy* 4340 p.p. *Fernandez* 437, 1490, *Lee Oldfield* 12 BM; HADHRAMAUT, *Popov, Tillin & Gilliland* 4134; A.-E. SUDAN, Kordofan (1842) *Kotschy* 28, Khartoum (1868) *Schweinfurth* 797 K, BM, Coast (1867) *Schweinfurth* 1486; ERITREA, *Beccari* 104, *Pappi* 1220 BM; FR. SOMALILAND, *Shantz* 5; BR. SOMALILAND, *Collenette* 96, *Gillett* 4553, *McKinnon* S/190, *Bally* 3772, *Peck* 286; SOCOTRA, *Balfour* 22 p.p.

C.3.m.18. **I. coerulea** *Roxb.*, Hort. Beng., 57 (1814) (nomen); Fl. Ind. **3**, 377 (1832).

"*I. argentea* L." sensu Bak., F.T.A. **2**, 97 p.p.; Fl. Brit. Ind. **2**, 98 p.p.; Fl. Mauritius and Seychelles 70 p.p., non L.

"*I. articulata* Gouan" sensu Prain & Bak.f., J. Bot. 1902, 141; Bak.f., L.T.A. No. 207; Andrews, Fl. Pl. A.-E. Sudan **2**, 216 p.p., non Gouan.

α. var. **coerulea.**

I. argentea L. var. *coerulea* (Roxb.) Bak., Fl. Brit. Ind. **2**, 99 (1879).

I. argentea L. var. *brachycarpa* Vatke, Oest. Bot. Zeit. **1878,** 213.

SAUDI ARABIA, Jeddah, *Zohrab* 293, *Trott* 1399, *Vesey Fitzgerald* s.n. BM, S. W. Rub el Kali *Thesiger* s.n. BM; YEMEN, *Lundt* 35; ADEN, *Balfour* s.n. *Perry* s.n. *Maxwell Darling* 194; HADHRAMAUT, *Wakefield expdn.* 29, *Guichard* 11 EA, BM; BALUCHISTAN; SIND; RAJPUTANA; BOMBAY PROV.; MADRAS PROV. without locality Roxb. Ic. 388 (ined.) K, *Roxburgh* (1813) s.n. BM syntypes; CEYLON; A.-E. SUDAN, Jebel Elba, *Newberry* 112 BM 119 BM; BR. SOMALILAND, Berbera, *Mrs. Lort Phillips* and *Bland* s.n. BM, Lasgori *Hildebrandt* 837 BM isotype of var. *brachycarpa.*

β. var. **occidentalis** *Gillett and Ali* var. nov.

I. argentea L. var. *polyphylla* Hochst. ex Defl., Voy. Yemen 129 (1889) (nomen.).

A var. *coerulea* foliolis superne strigulosis differt.

SAUDI ARABIA, Jeddah, *Zohrab* 308, Fatme valley *Fischer* 80, *Schimper* (1837) 970 BM, Sabya *Philby* 109 BM, Darb. 17°45′N, 42°15′E, *Vesey FitzGerald* 16416 BM; YEMEN, *Deflers* 130, *Schweinfurth* 1211, *Forskal* s.n. BM (labelled *I. tinctoria*) *Forskal* s.n. BM (labelled *I. spicata*); HADHRAMAUT, *Lunt* 173 BM, Dhofar *Bent* 9, 103; SIND, *Stocks* 488; FR. NIGER COL., Aïr 17°15′N, *Chopard & Villers* s.n.; IFAN; A.-E. SUDAN, *Kotschy* 29, 217 K, holo., BM isotype (syntype of var. *polyphylla*), *Schweinfurth* (1865) 1782, (1867) s.n. BM, (1868) 269 K, BM, *Pfund* 94, 345, *Broun* 1484, *Jackson* 2831, 3248 *B. Beshir* 155, *Dandy* 250 BM, *Massey* s.n. BM, *MacDougal* 32 BM; ERITREA, *Pappi* 3912 K, BM, 5167 BM, 6365 BM, 6468 BM, EA, *Schweinfurth & Riva* (1892) 358, *Bally* 6744, 6862; ETHIOPIA, Tigre, *Schimper* (1844) 1017 BM, 1854 syntypes of var. *polyphylla*, Roth 32, Schebelli *Donaldson Smith* s.n. BM; BR. SOMALILAND, *Gillett* 4171, 4930, *Glover & Gilliland* 345 K, BM, *Hemming* 144 EA, *Bally* 3756, 10158; SOMALIA, *Bally* 3756 EA; SOCOTRA, *Balfour* 22 p.p.; KENYA, K1, *Bally* 3450, 5482, 9087, *Gillett* 13321; K2 *Pole Evans and Erens* 1604; MAURITIUS (perhaps introduced) *Bojer* s.n., *Ayres* s.n. *Bouton* s.n.

Although the difference between the two varieties of *I. coerulea* is slight it seems worthwhile giving a varietal name to mark the difference because of the different areas of distribution of the two varieties; Arabia and the Red Sea coast of Africa forming a zone of overlap.

I am indebted to Mr. S. I. Ali of Karachi, now engaged on a revision of the *Papilionaceae* of W. Pakistan, for pointing out to me the clear and constant correlation between the breadth of suture, shape of fruit segment and number of leaflets which together, in our opinion, justify the specific distinction of *I. coerulea* from *I. articulata*. Prain and Bak.f., J. Bot. **40,** 141 (1902) recognise the three taxa here dealt with, though treating them all as forms of *I. articulata*. However, they state that *I. spicata* Forsk. is a synonym for *I. articulata* and that *I. houer* Forsk. refers to the Arabian plant intermediate between *I. articulata* s. str. and *I. coerulea*, i.e. to *I. coerulea* var. *occidentalis* as here accepted. There is indeed, as they say, a Forskal sheet in BM labelled (in an unknown hand) *I. spicata* which is actually *I. coerulea* var. *occidentalis*. But it does not at all agree with the description of *I. spicata*; e.g. the leaflets are not alternate, and *I. articulata* and *I. coerulea* are both of value for Indigo production while *I. spicata* is said to be useless for this purpose. The Copenhagen sheet of *I. spicata* does agree with the description and is conspecific with *I. hendecaphylla* Jacq.

The type of *I. houer* cannot be found. However, it is clear that the description does not apply to *I. coerulea* for Forskal confounded this plant with *I. articulata* under the name "*I. tinctoria*" and expressly contrasted his description of *I. houer* with that of his "*tinctoria*" as regards the narrower "cuneato oblongis" "semi-sex-jugis" leaflets and the "leguminibus rectis teretibus". All these four characters agree well with the African straight podded form of the true *I. tinctoria* L. which is indeed found in the Yemen and there can be but little doubt that this is the true *I. houer*. It is true that the Arabian name of *I. coerulea* does seem to be "Houer" or "Hawar" in contrast (according to Philby's m.s. notes in the Kew Herbarium) with "Adhlim" for *I. articulata* but this seems very slight evidence in comparison with Forskal's description. In the same paper Prain and Bak.f. demonstrate that *I. tinctoria* L. var. *brachycarpa* DC. is not, as was supposed in Fl. Brit. Ind. **2**, 99 a synonym for var. *coerulea* but an American plant and a form of *I. suffruticosa*.

Prain and Bak.f. also state that *I. cinerascens* DC, which had been treated as a synonym for *I. argentea* by Baker in his flora of Mauritius and the Seychelles, is conspecific with *I. tinctoria*. The elder Baker's opinion was based on a specimen at Kew collected by Bojer and labelled *I. cinerascens* which is definitely *I. coerulea* var. *occidentalis*. Thanks to the courtesy of Prof. Baehni in Geneva it has been possible to examine a photograph of De Candolle's type. This is a plant without fruit or mature flowers. Prof. Baehni says that the leaflets are glabrous above and that the indumentum is like that of *I. tinctoria* and not that of *I. articulata*. Prof. Baehni also says that the stamens are but 3 mm. long which would agree with *I. coerulea* rather than *I. tinctoria* where the stamens are nearly always 4 mm. long, but this could be due to the immaturity of the flowers. It is not quite impossible that De Candolle's plant could be conspecific with *I. coerulea* var. *coerulea* (which has not otherwise been seen from the Mascarenes) but on the whole it seems best to accept Prain and Bak.f.'s view that it is conspecific with *I. tinctoria* L.

C.3.m.19. **I. bogdanii** *Gillett* sp. nov.

α. var. **bogdanii.**

I. cryptanthae affinis, foliolis paucioribus differt.

Suffrutex ramis divaricatis appresse pilosis. *Folia* 5–9-foliolata, ubique pilis appressis densis vel subsparsis obtecta, pagina superiore glabra foliolorum sola excepta; stipulae deltoideo-subulatae, brunneae, *c.*1 mm. longae; rhachis 10–15 mm. longa, petiolo *c.*2 mm. longo incluso, saepius ultra foliola lateralia haud, vel minime, prolongata; petioluli 0·7 mm. longi; foliola suboblonga, in sicco viridia, basin paulo attenuata, apice apiculata, ad 5 mm. longa et 3 mm. lata. *Racemi* sessiles *c.*10–12-flori, *c.*10 mm. longi, post anthesin haud elongati, rhachi argentea, bracteis ovatis, *c.*0·4 mm. longis, deciduis, vel persistentibus, pedicellis argenteis, *c.*1·0 mm. longis, post anthesin valde reflexis. *Calyx* strigulosus, tubo *c.*0·8 mm. longo, lobis subaequalibus late deltoideo-acuminatis, *c.*0·4 mm. longis, sinu superiore quam sinibus aliis fere duplo latiore. *Vexillum* extus breviter appresse strigosum, ellipticum, apice emarginatum, breviter apiculatum, 3·4 mm. latum, 4·3 mm. longum. *Alae* glabrae, suboblongae, 3·8 mm. longae, unguiculo *c.*0·2 mm. longo incluso, basin 1 mm. latae, apicem versus 1·5 mm. latae, apice rotundatae. *Carina* extus appresse strigosa, 0·9 mm. calcarata, haud rostrata, 4 mm. longa.

Stamina omnia fertilia, filamentis 3–4 mm. longis, antheris anguste ovatis, basin muticis, *c.*0·7 mm. longis, appendice apicale rubro *c.*0·1 mm. longo incluso. *Ovarium* dense appresseque strigosum, 3–4-ovulatum, 3 mm. longum, in stylum glabrum 1·5 mm. longum abrupte rectangulare curvatum, stigmate globoso minuto. *Legumen* ut videtur indehiscens, diu persistens, brunneum, subsparse strigulosum, rectum vel paulo curvatum, sutura superiore quam inferiorem duplo latiore (*c.*0·6 et 0·3 mm.), 1–4-spermum, ad 12 mm. longum, valde torulosum, ad semina 2·2 mm. latum, 2·2 mm. crassum, inter semina 1·3 mm. latum, 0·9 mm crassum, endocarpio brunneo haud maculato. *Semina* laevia, nigra, subglobosa.

Kenya, K3, 11 km. from Eldama ravine on Kampi ya moto road, 1820 m., margin of road through *Acacia* thornbush, herb with branches radiating from central rootstock, annual?, in fruit only 25.9.1953, *Drummond & Hemsley* 4434 K, EA. Njoro-Rongai, 1950 m., plain, overgrazed pasture with *Pennisetum schimperi*, rare 3.6.1947, *Bogdan* 595. Elmenteita, Soysambu estate, plains with *Cynodon* and *Aristida adscensionis*, in rocky saline pan (temporary swamp), frequent, 31.7.1947, *Bogdan* 952 K, EA. Nr. Molo river on Nakuru-Eldama Ravine Rd., Esageri bush clearing scheme, 1650 m., cleared *Acacia-Tarchonanthus* forest, rare, 14.9.1948, *Bogdan* 2053. 24 km. N. of Thompson's falls, 2040 m., open rocky place, 5.9.1951, *Bogdan* 3227 holotype. Naivasha, 1800 m., standard green with pink and brown patches, wings pinks, keel greenish yellow, Aug. 1934, *Turner* 6772 K, EA.

Tanganyika, T1, Musoma district, Mom, 10.12.1956 in fruit, *Greenway* 9156.

β. var. **peteri** *Gillett* var. nov.

A var. *bogdanii* foliolis in sicco nigris, rhachi foliorum saepius ultra foliola lateralia bene producta, endocarpio paulo maculato, differt.

Tanganyika, T2 Ol Doinyo Samba west of Mt. Meru, 1600 m., 4.3.1914, *Peter* O.I. 54, K435. Southern edge Ngorongoro crater, 1780 m., shrublet 40 cm. tall, 24.7.1926, *Peter* 43210 holotype. Nr. Loliondo, 35°35′E, 2°5′S, 2100 m., open highland, sandy soil, low shrub 15–30 cm. tall browsed by game, 25.2.1954, *A. C. Brooks* 91.

C.3.m.20. **I. cryptantha** *Benth. ex Harv.*, Fl. Cap. **2**, 195 (1862) pro majore parte; N. E. Brown in Burtt Davy Man. Flor. Transvaal **2**, 371; L.T.A. No. 215.

α. var. **cryptantha.**

S. Rhodesia, Matopos, *Eyles* 7172, *Rattray* 94 BM,SRGH; Transvaal, *Burke* 449, *Zeyher* 473 isosyntypes, etc.; Orange Free State; N. Central Cape Prov.

Harvey quotes a further specimen *Dr. Sutherland* s.n. from Natal which is *I. arrecta* Hochst. ex A. Rich, his description seems, however, wholly based on the Burke and Zeyher specimens.

β. var. **desmodioides** (*Bak.*) *Gillett* stat. nov.

I. desmodioides Bak., J. L. Soc. **22**, 463 (1887), non Kew Bull. **1894**, 331.

I. pityophila Drake, Hist. Pl. Madag. **1**, 156 (1902).

Madagascar, *Baron* 889 isotype of *I. pityophila*, 3965 holotype, 4166.

An earlier *I. desmodioides* Benth. was published only in synonymy and thus invalidly and does not invalidate *I. desmodioides* Bak. (1887) as Drake supposed.

γ. var. **occidentalis** *Bak.f.*, Vierteljahrschr. Nat. Ges. Zur. **49**, 184 (1904).

S.W. Africa (*Fleck* 449 Z, 610 Z, 827a Z, *Dinter* 40 Z syntypes), *Pearson* 3663, 4342, 4656, 8244, 8314, 9115, *Örtendahl* 446, *Schoenfeldes* 470, *Codd* 5817, *Bradfield* 162.

As the types have not been seen there is a slight doubt about this variety. E. G. Baker relies chiefly on the small number of leaflets in some forms of var. *occidentalis* to separate it from var. *cryptantha*. It seems however that this character is variable and that a much better character is to be found in the thicker, straighter fruits.

C.3.m.21. **I. amorphoides** *Jaub. & Spach*, Ill. Pl. Or. t. 483 (1856); L.T.A. No. 220.

I. abyssinica Hochst. in Schimp. Pl. Abyss. 2213 (nomen).

I. alta Schweinf., Fl. Aeth. 11 (1867).

YEMEN; ADEN; ERITREA; ETHIOPIA (nr. Awawa 1853, 900–1200 m., *Schimper* P holotype), *Schimper* (1854) 2213 isotype of *I. abyssinica* and *I. alta*; BRIT. SOMALILAND; KENYA, Kl.

C.3.m.22. **I. cavallii** *Chiov.*, Ann. Di Bot. **13**, 380 (1915); L.T.A. No. 213.

SOMALIA, Benadir, *Paoli* 922 bis FI holotype.

C.3.m.23. **I. arrecta** *Hochst. ex A. Rich.*, Tent. Fl. Ab. **1**, 184 (1847); Bak. F.T.A. **2**, 97; L.T.A. No. 214; Hutch. & Dalz. F.W.T.A. **1**, 392 fig. 146; F.C.B. **5**, 162 t. 12; non Benth. ex Harv., Fl. Cap. **2**, 183 (1862).

"*I. anil L.*" sensu Guill. & Perr., Fl. Seneg. 180 (1832) non L.

I. cryptantha Benth. ex Harv., Fl. Cap. **2**, 195 (1862) pro min. parte, quoad pl. natalensem.

I. umbonata Welw. ex Bak., F.T.A. **2**, 98 (1871).

I. scopa De Wild. & Th. Dur., Ann. Mus. Congo. Bot. Ser. **3**: **1**, 60 (1901); L.T.A. No. 216.

S. ARABIA; SENEGAL; GAMBIA; PORT. GUINEA; FR. GUINEA; FR. SUDAN, fide Tisserant; GOLD COAST; NIGERIA; CAMEROONS; B. CONGO, *De Wevre* 1027 b BR holotype of *I. scopa*; A.-E. SUDAN; ERITREA; ETHIOPIA, *Schimper* (1837) 349 (1842) 1923 b isosyntypes; BRIT. SOMALILAND; UGANDA, U1–4; KENYA, K1–7; TANGANYIKA, T1–3, T5, T6 *Schlieben* 3830 BM, T7; PORT. E. AFR.; NYASALAND; S. RHODESIA; N. RHODESIA; ANGOLA, *Welwitsch* 2083 BM isotype of *I. umbonata*; MADAGASCAR; TRANSVAAL; NATAL, *Sutherland* s.n. syntype of *I. cryptantha*; E. CAPE PROV.

The range of this species has probably been greatly extended through cultivation, as it is one of the chief Indigo-producing species in Africa. It seems that hybridisation occurs between *I. arrecta* and *I. tinctoria*. In W. Africa, especially, plants occur which might be assigned to either species. In East Africa the two species are for the most part kept separate by the fact that *I. arrecta* occurs mainly at higher altitudes, *I. tinctoria* in the lowlands. It seems doubtful whether Dr. Cronquist is correct in treating *I. kisantuensis* De Wild & T. Dur. Bull. Herb. Boiss. **2**: **1**, 13 (1900); L.T.A. No. 219 as a synonym for *I. arrecta*. The fruits in the type, *Gillet* 839, are stipitate.

C.3.m.24. **I. longeracemosa** *Boiv. ex Baill.*, Bull. Soc. Linn. Par. **1**, 399 (1883); L.T.A. No. 217.

TRAVANCORE, *Barber* 6666, 6701, 6717, 6719; KENYA, K7 *Swynnerton* 281; ZANZIBAR, *Hildebrandt* 941, *Bojer*, s.n., *Greenway* 1263, *Vaughan* 1098, 1879; MADAGASCAR, *Boivin* s.n. P holotype, *Pervillé* s.n. BM.

C.3.m.25. **I. suffruticosa** *Mill.*, Gard. Dict. Ed. **8**, No. 2 (1768); L.T.A. No. 212; F.C.B. **5**, 162.

I. anil L., Mant. **2**, 272 (1771) fide Prain & Bak.f., J. Bot. **1902,** 137.

I. uncinata G. Don., Gard. Dict. **2**, 208 (1832).

I. micrantha Desv., Ann. Sci. Nat. Ser. **1**, vol. **9**, 410 (1826) fide Hook.f., non Bunge.

I. angolensis D. Dietr., Syn. Pl. **4**, 1036 (1847).

? *I. hofmanniana* Schinz., Verh. Bot. Ver. Brand. **30**, 162 (1888)? e descr.

I. suffruticosa Mill. var. *uncinata* Berhaut, Fl. Seneg. 50 (1954) (nomen).

INDIA (type of *I. anil*); PHILIPPINES; INDONESIA; C. VERDE ISLES; SENEGAL, fide Tisserant; GAMBIA; SIERRA LEONE, *unknown coll.* s.n. BM holotype of *I. uncinata*; LIBERIA; IVORY COAST, fide Tisserant; NIGERIA; ST. THOMAS; GABOON; B. CONGO; S. RHODESIA; ANGOLA (? type of *I. micrantha* Desv. = *I. angolensis* D. Dietr.); MADAGASCAR; S.W. AFRICA. Amboland (*Schinz* Z holotype of *I. hofmanniana*); JAMAICA. *unknown coll.* s.n. ex Miller herb. BM holotype.

This species which has been widely cultivated is a native of tropical America and is probably introduced in the Old World. In South Tropical Africa plants occur which are intermediate between this species and the next.

C.3.m.26. **I. tinctoria** *L.*, Sp. Pl. 751 (1753); L.T.A. No. 218.

I. houer Forsk., Fl. Aeg. Arab. 137 (1775).

I. sumatrana Gaertn., Fruct. **2**, 317 t. 148 (1791).

I. cinerascens DC., Prod. **2**, 226 (1825).

I. ornithopodioides Schum., Beskr. Guin. Pl. 372 (1829).

I. anil L. var. *orthocarpa* DC., Prod. **2**, 225 (1825).

I. orthocarpa (DC.) Berg in Berg & Schmidt Darstell. & Beschreib. offic. Gew. **4**, 30d (1863); Baker, F.T.A. **2**, 99 (1871) non *I. orthocarpa* Presl (1844).

I. bergii Vatke, App. Ind. Sem. Berol. 3 (1876).

I. tinctoria L. var. *torulosa* Bak.f., L.T.A. 157 (1926).

YEMEN (*Forskal* s.n. holotype of *I. houer*); INDIA (*Labillardière* s.n. G syntype of *I. anil* var. *orthocarpa* and *I. bergii*); E. PAKISTAN; CEYLON, *Hermann* s.n. Vol. 3, fol. 20 BM holotype; MALAYA; INDO CHINA; PHILLIPINES; INDONESIA (type of *I. sumatrana*); NEW GUINEA; C. VERDE ISLANDS; SENEGAL; FR. GUINEA; FR. SUDAN; GOLD COAST (*Thonning* s.n. C. holotype of *I. ornithopodioides*); TOGOLAND; *Warnecke* 262 holotype of var. *torulosa*; NIGERIA; OUBANGUI CHARI; A.-E. SUDAN; SOMALIA; SOCOTRA; KENYA, K7; TANGANYIKA, T3, T5, T6; ZANZIBAR; PEMBA; PORT. E. AFR.; N. RHODESIA; ANGOLA; MADAGASCAR (*unknown coll.* G syntype of *I. anil* var. *orthocarpa* and *I. bergii*) MAURITIUS, or Reunion, *unknown coll.* (G) K photo, type of *I. cinerascens*; TROP. AMERICA (introduced).

I. tinctoria as here delimited shows much variation: it is not unlikely that two or more taxa, originally distinguishable in the wild state, have become confused in cultivation. The confusion seems now too great for any attempt at infraspecific classification by ordinary taxonomic methods, without cultivation and breeding experiments, to succeed. Berg, Baker, Taubert and others have separated a wild African form with short inflorescences as *I. bergii* (*I. orthocarpa*) from the more luxuriant true *I. tinctoria*. The two forms seem however to blend into one another. Incidentally Prain and Bak.f. who examined the syntypes of *I. anil* var. *orthocarpa* DC. from India and Madgascar state (J. Bot. **40**, 140 (1902)) that they represent two cultivated forms of *I. tinctoria* neither of which is the same as the African plant to which the name was later applied.

There has been some doubt about the identity of *I. houer* Forsk. of which the type is lost. Schweinfurth (Bull. Herb. Boiss **4**, app. **2** (1896) 242) suggested that it might be the same as *I. arrecta* Hochst. Schwarz (Fl. Trop. Arabia (1939) 104) treats it as a synonym for *I. tinctoria* L. but without giving reasons. The description would fit either species. Forskal's locality Surdud can readily be found on modern maps and all places in the area are well below 100 m. *I. arrecta* in E. Trop. Africa and Arabia is usually found at higher altitudes, not being recorded below 2000 m. in Arabia while *I. tinctoria* occurs usually, if not invariably, below 1000 m. There thus seems a very high degree of probability that Forskal's plant is *I. tinctoria*. Gamble in the Flora of Madras (1918) and Rydberg in the N. American Flora (1923) separate *I. sumatrana* from *I. tinctoria* on the grounds of its somewhat thicker pods, which Rydberg further states to be more curved and correlated with leaflets somewhat strigose above instead of glabrous. However, there seem to be little or no correlations between Rydberg's three characters and intermediate pod thicknesses are frequent. It therefore seem better to follow previous workers on *Indigofera* in Africa and not to attempt to separate these forms. The slender torulose fruits of *Warnecke* 262 the type of var. *torulosa* Bak.f. seem at first sight very different from those of *I. tinctoria*. However several sheets from Nigeria, e.g. *Barter* s.n., *Dalziel* 58 show such fruits on the same plant as normal fruits and it seems that they may be simply a stage in the development of the latter.

C.3.m.27. **I. ciferrii** *Chiov.*, Att. Ist Bot. Univ. Pavia Ser. **4**: **7**, 128, 4.4 (1936).

SOMALIA, lower Webi Shebeli *Ciferri* 63 bis (FI holo.) K isotype.

A little understood species of which more material is desired.

C.3.m.28. **I. sparteola** *Chiov.*, Fl. Somala **1**, 140, t. 13 (1929); L.T.A. p. 881.

SOMALIA, nr. Obbia *Puccioni & Stefanini* 397 FI, 436 FI syntypes. *Drake-Brockman* 955.

C.3.m.29. **I. pungens** *E. Mey*, Comm. Pl. Afr. Austr. 93 (1836); Fl. Cap. **2**, 170.

S.W. AFR., *Dinter* 8168, *Galpin* 14146, *Acocks* 15668, *Pearson* 3621; N.W. CAPE PROV., Little Namaqualand, *Pillans* 5373 (*Drège* s.n. type).

C.3.m.30. **I. cliffordiana** *Gillett* sp. nov.

*Suffrutex c.*50 cm. altus, ramis dense appresseque argenteis. *Folia* normaliter 3-foliolata; stipulae subulatae *c.*2 mm. longae; rhachis haud vel 1–2 mm. ultra foliola lateralia prolongata; petiolus 2–3 mm. longus; foliola cuneato-obovata, apice apiculata, nervo mediano superne impresso, utrinque pilis rigidis appressis vestita, *c.*12 mm. longa, *c.*7 mm. lata. *Racemi* 10–20-flori, 10–30 mm. longi, pedunculo 2–12 mm. longo incluso, rhachide argentea, bracteis lanceolatis mox deciduis *c.*1·5 mm. longis, pedicellis *c.*1·5 mm. longis, post anthesin valde reflexis. *Calyx* strigosus, tubo *c.*1 mm. longo, lobis subaequalibus deltoideo-acuminatis *c.*2 mm. longis, sinu superiore quam alios duplo latiore. *Vexillum* extus strigosum, suborbiculare, 4·5 mm. longum, 4·5 mm. latum. *Alae* glabrae dolabriformes, 4·3 mm. longae, unguiculo 0·2 mm. longo incluso, ad 1·5 mm. latae. *Carina* externe, praecipue ad commissuram, strigosa, *c.*0·7 mm. calcarata, haud rostrata, *c.*4·3 mm. longa. *Stamina* omnia fertilia, filamentis 3·5–4 mm. longis, libero breviore *c.*3 mm. longo, antheris ellipticis, basin muticis, 0·6–0·8 mm. longis, appendice apicale 0·05–0·2 mm. longo incluso. *Ovarium* 12-ovulatum, appresse strigosum, 3 mm. longum, stylo glabro 1 mm. longo ad basin abrupte curvato, stigmate globoso *c.*0·1 mm. diametro.

[*Legumen* argenteum valde arcuatum, inter semina complanatum, ad 20 mm. longum, endocarpio haud maculato. *Semina* (haud matura) 6–8, oblonga, *c.*1·8 mm. longa, 0·9 mm. lata, 0·7 mm. crassa.]

KENYA, K1 Furroli 3°42′N, 38°0′E, *c.*950 m., undulating lava plateau, transition from *Commiphora-Acacia* open scrub to subdesert with scattered *Acacia spirocarpa*, rainfall probably *c.*250 mm., *c.*50 cm. tall, corolla brick red, locally common, 12.9.1952, *Gillett* 13804 K holo. EA isotype. *Gillett* 13804A (a variant with some leaves 4- or 5-foliolate). Uaso Nyiro, near Archers Post, 900 m., flowers pink, *Mrs. Dalton* 3 EA. 32 km. S.E. of Lodwar, on sand dunes, resembling *I. spinosa* but without spines, 18.7.1954, *C. F. Hemming* 338 K, EA. K2, Lorugumu, 600 m., sandy places, woody herb 30–50 cm. tall. flowers red, 12.4.1954, *G. Popov* 1482 BM. K6, 72 km. from Nairobi on Magadi road, 1050 m., semi-desert grass and shrublets on white rocky soil, shrublet 30–50 cm. tall, flowers red, 26.6.1952 in fruit, *Bogdan* 3470.

NOTE: *Bogdan* 3470 (from which the fruit is described) and *Mrs. Dalton* 3 differ from the type in having longer petioles (up to 5 mm.) and a rhachis often prolonged well (up to 3 or 4 mm.) beyond the lateral leaflets.

I have much pleasure in naming this species after Col. E. H. M. Clifford, O.B.E., M.C., senior British Commissioner on the British Somaliland-Ethiopia and Kenya-Ethiopia boundary commissions, to whose kindly enthusiasm for scientific research the attachment to these commissions of botanists and zoologists, the provision of facilitities for their work, and hence the discovery of this and many other species, have been due.

C.3.m.31. **I. dimidiata** *Vog. ex Walp.*, Linnaea **13**, 521 (1839); Fl. Cap. **2**, 176, including var. *laxior* Bak.f., J.Bot. **58**, 76 (1920); L.T.A. No. 48.

PORT. E. AFR Tsetserra, 2220 m., *Wild* 4477 K, SRGH; NYASALAND, Mlanje, *Mrs. Shinn* s.n. BM; S. RHODESIA, Umtali, Engwa, 2100 m., *Exell, Mendonça & Wild* 307 BM; TRANSVAAL, *Rogers* 21076 K, BM type of var. *laxior*; NATAL; BASUTOLAND; E. CAPE PROV. (*Ecklon & Zeyher* 1852 p.p. type).

This is very distinct from all other Tropical African species. It is one of the very few members of subgenus **Indigofera** in which the corolla is almost glabrous. The keel is spurred and not rostrate and the stamens are normal, thus showing no affinity with **Indigastrum** and **Microcharis**. It is placed by Harvey along with some related S. African species in a section **Trifoliolatae** in which the leaves are digitately trifoliolate. There does not seem, however to be any special affinity between the various tropical African plants which possess this characteristic and the section is therefore not recognised here.

C.3.m.32. **I. repens** *Cronquist*, Bull. J. Bot. et Brux. **22**, 225 (1952); F.C.B. **5**, 164.

B. CONGO, *Taton* 273, (645 BR holo.).

Subsect. C.3.n. **Hirsutae** *Rydberg*, N. Amer. Flor. **24**, 140 (1923).

Type species *I. hirsuta* L.

Centre of distribution Angola.

Leaves pinnate, the rhachis prolonged beyond the lateral leaflets. Hairs simple or with very unequal arms, often long. Calyx lobes much longer than the tube almost or quite as long as the stamens. Indumentum of standard not shiny and closely appressed. Alae pilose, at least at the margins. Pod straight, not long, rather stout, deflexed clothed with rather long, often brown, hairs.

Spreading hairs on calyx and fruit conspicuous, more than 0·8 mm. long; leaflets usually more than 5:

Leaflets rarely more than 18 mm. long or 10 mm. wide; petiole often under 4 mm. long:

Peduncle usually longer than the dense inflorescence; ovules *c*.3; seeds oblong-rounded, minutely dotted, not coarsely pitted; endocarp dark, not spotted; pods usually under 8 mm. long, *c*.2 mm. wide; hairs on calyx dark brown or black . . . . 1. *longebarbata.*

Peduncle usually shorter than the lax, slender inflorescence; hairs on young fruit mainly white . . . . 2β. *hirsuta var. pumila.*

Leaflets often more than 18 mm. long and 10 mm. wide; petiole hardly ever less than 4 mm. long; peduncle usually shorter than inflorescence; pods over 8 mm. long; endocarp strongly spotted; seeds angular, coarsely pitted:

Peduncle well developed, at least twice as long as a fruit; fruits usually 6-seeded, *c.*2 mm. wide and *c.*7 times as long; some of the hairs on dorsal side of fruits often brown; leaflets 5–7, rarely 9 . . . . 2α. *hirsuta var. hirsuta.*

Peduncle short, rarely twice as long as a fruit; fruits usually 2–4-, sometimes 5–6-seeded, *c.*2·5–3 mm. wide, less than 7 times as long, ± tetragonal; hairs on dorsal side of fruit usually mostly white; leaflets usually up to 9, sometimes up to 11 or 13 · · · · · · · · · · · · · · · · 3. *astragalina.*

Hairs on calyx and fruit much less conspicuous, 0·5 mm. long, or less, those on the fruit white; fruit 1–1·5 mm. wide; endocarp spotted; leaflets 3–5 . . . . . . 4. *deightonii.*

C.3.n.1. **I. longebarbata** *Engl.*, Hochgebirgsflora Afrikas 257 (1892); L.T.A. No. 177; F.C.B. **5**, 141.

"*I. hirsuta L.*" sensu Harv., Fl. Cap. **2**, 194 (1862) non L.

I. hirsuta L. var. *polystachya* Welw. ex Bak., F.T.A. **2**, 89 (1871).

I. schlechteri Bak.f., Viertel Jahrschr. Nat. Ges. Zur. **49**, 192 (1904).

I. griquana Schltr. ex A. Zahlbr., Ann. Nat. Hof. mus. Wien **20**, 23 (1905).

NIGERIA, Bauchi Plateau, *Lely* 510, 557, *Dent Young* 45; BRIT. CAMEROONS, Bamenda, *Johnstone* 221/31 FHI; BELG. CONGO, east and south; ETHIOPIA, *Schimper* (1863) 1389 isotype; UGANDA, U2; KENYA, K3–5; TANGANYIKA, T2, T4, T6–7; PORT. E. AFR., Niassa distr. Vila Cabal; NYASALAND; N. RHODESIA; ANGOLA, *Welwitsch* 2036, 2056 isosyntypes of *I. hirsuta* var. *polystachya*; TRANSVAAL; SWAZILAND; ORANGE FREE STATE; NATAL (*Schlechter* 6499 syntype of *I. schlechteri* and of *I. griquana*), *Sutherland* s.n. (*I. hirsuta* sensu Harv.); E. CAPE PROV., *Tyson* 1273 syntype of *I. schlechteri*, (*Penther* 2645 syntype of *I. griquana*).

This species is hardly found below *c.*1500 m. in Tropical Africa and goes up to 2500 m., which is high for the genus. It is variable, plants from Angola and N. Rhodesia for example have peduncles shorter than usual while S. African plants tend to have unusually small and numerous leaflets.

C.3.n.2. **I. hirsuta** *L.*, Sp. Pl. 751 (1753); L.T.A. No. 173; F.C.B. **5**, 140 t. 10.

α. var. **hirsuta.**

? *I. lateritia* Willd., Sp. Pl. **3**, 1233 (1803)?

I. ferruginea Schum. & Thonn., Beskr. Guin. Pl. 370 (1829).

I. fusca G. Don., Gard. Dict. **2**, 211 (1832).

S. CHINA; BENGAL; BOMBAY; MADRAS; BURMA; CEYLON, *Hermann* 172 BM holotype; INDOCHINA; PHILLIPINES; NEW GUINEA; N. AUSTRALIA; C. VERDE ISLES; SENEGAL; GAMBIA; FR. GUINEA; SIERRA LEONE; LIBERIA; FR. SUDAN; GOLD COAST (*Thonning* s.n. C. holotype of *I. ferruginea*); TOGO; DAHOMEY; NIGERIA; CAMEROONS; ST. THOMÉ, *G. Don* s.n. BM type of *I. fusca*; GABOON; FR. CONGO; OUBANGUI; B. CONGO; A.-E. SUDAN; UGANDA, U1–4; KENYA, K7; TANGANYIKA, T1, 3, 4, 6, 8; ZANZIBAR; PEMBA; PORT. E. AFR.; S. RHODESIA; N. RHODESIA; ANGOLA; TROP. BECHUANALAND; MADAGASCAR; TROP. AMERICA (probably introduced).

A weed of cultivation whose range has certainly been increased by man.

β. var. **pumila** *Welw. ex Bak.*, F.T.A. **2**, 89 (1871).

NIGERIA, Bauchi plateau *Batten Poole* in FHI 13187 FHI, *Lely* P720; ANGOLA, Huilla, in more elevated pastures, *Welwitsch* 2057 isotype. Ganda 269 km. inland 12°S 1350 m. *Mrs. Faulkner* A6; Membranoco Cubae 1050 m. *Mrs. Faulkner* A189.

This plant needs further investigation. Mature fruits are unknown. It is, in some respects, intermediate between *I. hirsuta* and *I. longebarbata* and would perhaps be better placed as a variety of the latter species.

The identity of *I. lateritia* Willd. is obscure. Firstly it is not clear whether Wildenow's description is based on a specimen, "*I. hirsuta L.*" sensu Jacq. coll. Bot. **2**, 359 and Ic. Pl. Rav. 3 t. 569 being merely cited as a synonym, or whether Jacquin's description and plate are to be regarded as the type. Prof. E. Werdermann of Berlin kindly searched Willdenow's herbarium for a type, but found none. He writes that the folder labelled *I. lateritia* has been annotated by Willdenow personally as *I. viscosa* (*I. suaveolens*) and considers that Willdenow himself discarded his own species as being synonymous with *I. viscosa* Lam. This opinion is adopted by J. G. Baker in F.T.A. **2**, 82 and Fl. Br. Ind. **2**, 95 and by E. G. Baker in J. Bot. **41**, 242 (1903). On the other hand E. G. Baker in L.T.A. 123 and 146 (1926) thinks that *I. lateritia* is more probably a synonym for *I. hirsuta*. Neither Willdenow's description nor Jacquin's description and plate really fits either of these two species, nor yet any other species known to me. Perhaps, on the whole, *I. lateritia* is best treated as a *nomen ambiguum*.

C.3.n.3. **I. astragalina** *DC.*, Prod. **2**, 228 (1825); L.T.A. No. 176; F.C.B. **5**, 141 in obs.

"*I. hirsuta L.*" sensu Bak., Fl. Brit. Ind. **2**, 98 (1879) p.p. non L.

E. PUNJAB; SIKKIM; CENTRAL INDIA; BENGAL; BOMBAY; MADRAS; CEYLON; SENEGAL, *Perrottet* (P or G holo); FR. SUDAN; FR. NIGER COL.; N. NIGERIA; BELG. CONGO, East, *Hendrickx* 4414 EA; A.-E. SUDAN; ERITREA; KENYA K2; TANGANYIKA, T1, T4–8; PORT E. AFR.; NYASALAND; S. RHODESIA; N. RHODESIA; ANGOLA; S.W. AFRICA; TRANSVAAL; W. INDIES (introduced and apparently scarce).

It is remarkable that, while botanists dealing with African plants have, since 1825, more or less consistently distinguished *I. astragalina* from *I. hirsuta*, those dealing with Indian plants have, until now, confused them. This may be because De Candolle's type seems to belong to an unusually few-seeded form of the species which may not exist in India. The two species are certainly closely related and it seems that introgressive hybridisation may take place in areas where both occur. For example many southern Rhodesian specimens are more or less intermediate between the two species. *I. astragalina* is apparently the more xerophytic of the two species, as appears from the distributions summarised above and from occasional collectors' notes. Thus from Bombay Province Sedgewick and Bell record "rainfall $\pm$ 750 mm." for their 5333 (*I. astragalina*) but "rainfall 3000 mm." for their 6732 (*I. hirsuta*).

C.3.n.4. **I. deightonii** *Gillett*, Kew Bull **1955**, 580 (1956).

FR. GUINEA; SIERRA LEONE; LIBERIA; NIGERIA; A.-E. SUDAN.

Subsect. C.3.o. **Microcarpae** *Rydberg*, N. Amer. Flor. **24**, 141 (1923).

Type species *I. domingensis* Spreng.

Centre of distribution, in America.

Leaflets 5–9, punctate beneath. Calyx lobes subulate, 2–3 times as long as the tube. Fruit short, reflexed, subtorulose. Endocarp spotted. Seeds nearly cubic with rounded edges.

C.3.o.1. **I. microcarpa** *Desv.*, J. Bot. 3, 79 (1814); L.T.A. No. 205.

I. perrottetii DC., Prod. **2**, 228 (1825).

"*I. enneaphylla L.*" sensu Bak., F.T.A. **2**, 95 (1871), non L.

I. richardiana Baill., Bull. Soc. Linn. Par. **1**, 398 (1883).

I. boiviniana Baill., ibid.

SENEGAL, *Perrottet* s.n. (P holo.) K, BM isotypes of *I. perrottetii*; FR. SUDAN; GOLD COAST; FR. NIGER COL.; N. NIGERIA; KENYA, K7; TANGANYIKA, T4–5; PORT. E. AFR.; S. RHODESIA; N. RHODESIA; ANGOLA; MADAGASCAR (*Boivin* 2439 P type of *I. boiviniana, Richard* 139 P type of *I. richardiana*), etc.; MEXICO; E. TROP. S. AMERICA (*unknown collector* P. holotype); W. TROP. S. AMERICA.

The fruit is rather variable. The species seems to be spreading in Africa in cultivated ground, and along river banks. It is perhaps indigenous in South America only. It seems doubtful whether two species which have been segregated from *I. microcarpa* in America; *I. domingensis* Spreng ex DC., Prod. **2**, 227, and *I. sabulicola* Benth. in Mart. Fl. Bras. **15**, 40 (1859) are truly distinct. African specimens tend to be a good deal stouter than those from America but otherwise are not distinguishable.

Subsect.C.3.p. **Alternifoliolae** (*Harvey*) *Gillett*.

"Group" *Alternifoliolae* (*Alternifoliae* per errore) Harvey, Fl. Cap. **2**, 167 (1862).

Type species *I. spicata* Forsk.

Centres of distribution in Angola and the A.-E. Sudan.

Leaves imparipinnate, the leaflets all, or mostly, alternate: sometimes leaflets 2 only, 1 terminal and 1 lateral. Rarely some of the leaves simple. No glandular or dark brown hairs present. Fruiting pedicels reflexed (except in No. 1). Endocarp frequently not spotted.

The following key includes all tropical African species with alternate leaflets whether members of this section or not. Further such species occur in Madagascar and Mexico.

Glandular multicellular hairs numerous, short . . . . *sordida.*
Glandular hairs absent:
  Fruits elongated, erect or spreading on pedicels 3–4 mm. long . . . 1. *malongensis.*
  Fruits reflexed, or, if erect, closely aggregated on short pedicels:
    Axis of inflorescence spinescent; argenteous desert shrublet; leaflets about 5, cuneate obovate, *c.*3 mm. long . . . . 2. *intricata.*
    Axis of inflorescence not spinescent:
      Pods hardly longer than wide, 1- or rarely 2-seeded; endocarp not spotted;
      Leaflets more than 3; pods straight; inflorescence *c.*10-flowered; stipules wide, brown, scarious . . . . 19. *kerstingii.*
      Leaflets 2–3; pod curved so that style is near base; inflorescence 2–4-flowered:
        Hairs on fruits stems and leaves closely appressed; leaflets oblong-lanceolate or linear, 2–10 times as long as wide, 3–15 mm. long; peduncle rarely less than 5 mm. long . . . . . 3. *nephrocarpoides.*
        Hairs on fruits, and often on stems and leaves also, crisped, spreading at the tips; leaflets obovate, elliptical or orbicular, less than twice as long as wide, 1·5–5 mm. long; peduncle less than 5 mm. long . . . 4. *nephrocarpa.*
      Pods much longer than wide, 2 or more seeded:
        Longer leaflets less than 3 cm., or, if that length, then calyx lobes hardly longer than tube:
          Leaflets never more than 2; pod not above 10 mm. long, densely clothed with long ± spreading hairs:
            Sterile tapering distal part of pod curved gradually upwards, raceme very dense, almost, or quite sessile . . . . 5. *diphylla.*

Sterile tapering part of pod in a straight line with the rest, only the
nearly glabrous style bent abruptly upwards; raceme laxer,
pedunculate. . . . . 6. *flavicans*.

Leaflets usually more than 2; pod often over 10 mm. long, its indumentum various, usually closely appressed:

Pods straight, or almost so, or, if curved (in No. 7), then calyx much
shorter than corolla:

Racemes very short, of 2–3 flowers, or flowers single in leaf axils;
intricate twiggy shrublet, leaflets 1–3, less than 1 cm. long . . .
ruspolii.

Racemes usually longer, of more than 3 flowers:

Calyx much less than half as long as corolla, the lobes not much
longer than the tube; standard densely covered outside with
closely appressed stiff silky hairs:

Leaflets 1–5, oblong *c*.2–4 times as long as wide, the terminal
often much longer than the rest; pod usually ± torulose
and curved; mature inflorescence usually under 11,
sometimes up to 15 cm. long . . . . 7. *oblongifolia*.

Leaflets 3–11, obovate or elliptical, *c*.1·5–3 times as long as
wide, the terminal as long as, or but slightly longer than,
the rest; pods rarely torulose or curved; mature inflorescence usually over 11, sometimes up to 22 cm. long . . .
8. *schimperi*.

Length of leaflets rarely less than 10 mm.:

Hairs on stems straight, closely appressed:

Stamens 4–7 mm. long; pod usually not bent at base
α. var. *schimperi*.

Stamens 7–10 mm. long; pod bent sharply at base . . .
β. var. *baukeana*.

Hairs on stems crisped, spreading at the tips; stamens
under 7 mm. . . . . γ. var. *crispidula*.

Length of leaflets 5 mm. or less; hairs straight, appressed;
stamens under 7 mm. . . . . δ. var. *parvifoliolata*.

Calyx at least half as long as corolla, the lobes usually much
longer than the tube; hairs on standard relatively sparse,
often ± spreading:

Pods 5 mm. long or less, or, if up to 10 mm. (in *I. sessiliflora*),
then the inflorescence a compact axillary cluster:

Inflorescence 2 or more times as long as subtending leaf;
pod torulose, 2-seeded, endocarp not spotted, seeds not
pitted; leaflets *c*.5 on a very short rhachis, the whole
plant densely covered with ± spreading silvery hairs
9. *semitrijuga*.

Inflorescence a dense sessile axillary cluster shorter than the
subtending leaf, or (in *I. enneaphylla*) sometimes
pedunculate and as long as the leaf:

Stipules ± triangular, brown scarious, glabrescent; pod
c.2-seeded somewhat tetragonal; endocarp not
spotted, seeds not pitted . . . . *enneaphylla*.

Stipules narrow lanceolate, thickly strigose:

112

Pod *c.*2 mm. wide, somewhat tetragonal, not torulose,
2-seeded; endocarp spotted; seeds with 4 pits on
each face . . . . *trigonelloides.*

Pod *c.*1·5 mm. wide, not tetragonal, torulose, 3–6
seeded; endocarp not spotted; seeds not pitted
sessiliflora.

Pods over 5 mm. long; inflorescence a ± lax raceme:
Hairs (or many of them) on stems, peduncles and leaf
rhachises spreading at the tips, giving the plant a ±
hirsute appearance, hairs on upper leaf surface slender,
not very straight; endocarp not spotted:
Leaflets more than 8, pod blackish . . . .
10α. *daleoides* var. *daleoides.*

Leaflets fewer than 8:
Plant silvery; pod white; terminal leaflet about same
size as others . . . . 11. *maritima.*

Plant greenish; pod blackish; terminal leaflet longer
than others . . . . 12. *volkensii.*

Hairs on stems, peduncles and leaf rhachides all, or almost
all, closely appressed:
Leaflets rarely more than 4, definitely not glabrous above:
Hairs on upper surface very fine, more slender than
those beneath; leaflets ± elliptic; stems slender
(*c.*1 mm. diam.) pod 10 mm. long, or less . . . . .
13. *insularis.*

Hairs on upper leaf surface thicker, very straight and
closely appressed, much like those beneath:
Calyx teeth shorter than mature corolla:
Terminal leaflet ± narrow elliptic, lanceolate or
oblong, up to 30 mm. long; leaflets 2–5, if 4
then the middle 2 rarely opposite . . . .
14. *phillipsiae.*

Terminal leaflet obcordate-lanceolate up to 7 mm.
long; leaflets nearly always 4, the middle 2
opposite, the lowest solitary. 15. *sesquijuga.*

Calyx teeth longer than corolla . . . . 16. *viridiflora.*
Leaflets more than 4, or, if sometimes 3 (in forms of
17 and 18), then glabrous above:
Stipules with a wide scarious base which is glabrescent,
at least at the edges; hairs well separated,
*c.*0·5 mm. long:
Pod straight or sometimes slightly curved downwards,
often subtorulose when young; leaflets usually
more than 5, rarely glabrous above; peduncles
usually over 10 mm. long; endocarp not spotted;
seeds not pitted . . . . 17. *spicata.*

Pod usually slightly curved upwards, somewhat
flattened, not torulose; leaflets often 5 or fewer,
glabrous above; peduncle usually under 10 mm.
long; endocarp spotted; seeds pitted . . . . . .
18. *oxalidea.*

Stipules ± subulate and strigose:

Upper leaf surface visible between the relatively sparse hairs; hairs on lower leaf surface relatively large (*c*.1 mm. long); leaflets usually, 7–13, but fewer in seedlings; endocarp with large pale spots . . . . 22. *alternans*.

Upper leaf surface hidden by relatively dense hairs:

Leaflets 3–9 in number; peduncles not over 8 cm. long:

Length of leaflets under 1 cm. endocarp not spotted . . . . 23. *hololeuca*.

Length of at least the terminal leaflet usually over 1 cm.:

Calyx lobes triangular, *c*.1–2 times as long as tube; young and old leaflets white . . . 24. *benguellensis*.

Calyx lobes narrow subulate, 3 or more times as long as tube; young leaflets ± golden; endocarp with large pale spots . . . . . . 25. *auricoma*.

Leaflets 9–15 in number; peduncles up to 20 cm. long; endocarp not spotted . . . . 10β. *daleoides* var. *gossweileri*.

Pods curved; calyx lobes much longer than the tube:

All leaflets under 2 cm. long; endocarp ± spotted; seeds close together:

Pods circinnate, not flattened; stem prostrate not rooting at nodes . . . . 20. *circinella*.

Pods falcate, flattened; stem prostrate, rooting at nodes . . . . . 21. *radicifera*.

Longer leaflets over 2 cm. long; endocarp not spotted; seeds well separated; stems erect from a woody rootstock . . . . 26. *antunesiana*.

Longer leaflets over 3 cm. long; calyx lobes much longer than tube; lower leaves sometimes simple:

Leaflets rarely less than 4 mm. wide, or more than 10 times as long as wide:

Upper leaves having up to 11 leaflets, which are often subopposite, may be only 2–3 cm. long, and are rough on both surfaces with minute stiff hairs *c*.0·2 mm. long arising from pale spots; fruit coiled, endocarp not spotted; seeds well separated . . . . 26. *antunesiana*.

Upper leaves with rarely more than 5 leaflets which are usually glabrous and minutely dark punctate above, except at the margins; hairs, if present *c*.0·5 mm. long; fruits straight; endocarp spotted; seeds close together . . . . . . 27. *conjugata*.

Leaflets glabrous above, or with a few appressed hairs near the edge:

Stems, petioles and inflorescences with closely appressed hairs:

Leaflets of compound leaves *c*.4–14 times as long as wide . . . α. var. *conjugata*.

> Leaflets of compound leaves *c*.2·5–3 times as long as wide
> β. var. *trimorphophylla*.

> Stems petioles and inflorescences densely covered with crisped
> spreading hairs, leaves often simple . . γ. var. *occidentalis*.

> Leaflets appressed strigose all over the upper surface . . . . . .
> δ. var. *schweinfurthii*.

> Leaflets not above 2 mm. wide, *c*.20–30 times as long as wide, densely
> appressed strigose on both surfaces . . . . 28. *chevalieri*.

C.3.p.1. **I. malongensis** *Cronquist*, Bull. J. Bot. Et. Brux. **22,** 220 (1952); F.C.B. **5,** 147.

B. Congo, South, *Vin* 1 BR holotype.

With its erect fruits this species is out of place in this subsection. Its true affinities cannot be ascertained till more complete material is available.

C.3.p.2. **I. intricata** *Boiss.*, Fl., Or. **2,** 190 (1872); non sensu Balf. f., Bot. Socotra, Trans. R. Soc. Edin. **31,** 74 (1888), nec. sensu Hutch & Bruce, Kew Bull. **1941,** 118.

Arabia, Trucial Coast, *Mrs. Holmes* 355, *Guest* s.n.; Oman, *Fernande* 3249, 331; *Aucher-Eloy* 4342 isosyntype; S. Persia, *Aucher Eloy* 4343 isosyntype, *Bornmueller* 245, *Parsa* 91; Baluchistan, *Pierce* s.n.

The specimens from Socotra attributed to this species by Balfour are *I. pseudo-intricata*. Those from Somaliland quoted by Hutchinson and Bruce are *I. spinosa* (*Gillett* 3951) and sp. aff. *I. spiniflorae* (*Gillett* 4532).

C.3.p.3. **I. nephrocarpoides** *Gillett* sp. nov.

I. nephrocarpa Balf.f., Proc. Roy. Soc. Edin. **11,** 509 (1882); Trans. R. Soc. Edin. **31,** 63 (1888) pro min. parte.

Ab *I. nephrocarpa* ut in clavi distinguitur.

Herba perennis radice sublignoso, ramis tenuibus, flexuosis argenteis, pilis rigidis, arcte appressis, *c*.0·5 mm. longis, dense obtectis. *Folia* alterne 2–3-foliolata; stipulae subulatae albidae, *c*.1 mm. longae; petiolus argenteus (pilis ut ad ramos), 5–15 mm. longus (quam foliola saepius longior); foliola oblonga, lanceolata vel linearia, apice ± acuta, inferne dense, superne sparse, appresse strigosa, 3–15 mm. longa, 1·5–2·5 mm. lata, longitudine quam latitudine 2–10-plo. maiore. *Racemus* 2–4-florus, 2·5 mm. longus, quam pedunculus appresse argenteus 4–15 mm. longus multo brevior, bracteis *c*.0·7 mm. longis, pedicellis post anthesin reflexis, *c*.1·5 mm. longis. *Calyx* strigosus, tubo *c*.0·6 mm. longo, lobis deltoideis *c*.1·1 mm. longis. *Vexillum* extus strigosum late ovoideum, 2 mm. longum, 2 mm. latum. *Alae* glabrae, suboblongae, 0·7 mm. latae, 2·2 mm. longae, unguiculo 0·2 mm. longo incluso. *Carina* ad apicem paulo strigosa, 0·2 mm. obtuse calcarata, haud rostrata, 2·2 mm. longae. *Stamina* omnia fertilia, filamentis 1–1·5 mm., antheris ovoideis *c*.0·3 mm. longis, apice apiculatis, apice et ad basin pilis paucis rigidis obsitis. *Ovarium* 1-spermum, *c*.0·3 mm. longum; styli pars inferior recta strigosa 0·7 mm. longa, pars superior curvata glabra; stigma parvum, capitatum, incurvatum. *Legumen* reniforme, 1-spermum, argente strigosum, apice fere ad basin sursum inflexa, 2 mm. longum, 1·7 mm. latum, 0·8 mm. crassum. *Semen* brunneum, reniforme, 1·6 mm. longum, 1·0 mm. latum, 0·4 mm. crassum.

Socotra, plains, Feb.-March 1880, *Balfour* 104 K holo. BM isotype. *Aug.* 1880, *Balfour* 336. Reiged, 450 m., prostrate herb, flowers deep pink, 4.4.1953, *G. Popov.* SO/337 BM.

Prof. Balfour l.c. 1888 discusses this plant, which he considers but a form of
I. nephrocarpa due to environmental conditions. This view was plausible so long
as only stunted examples of *I. nephrocarpa* were known. However *Bent* 22 from
Dhofar, which is a luxuriant example of *I. nephrocarpa* with relatively large leaves,
still has the broad short leaflets, crisped indumentum short petioles and short
peduncles characteristic of that species and the same is true of *Popov* SO/148,
except that there the indumentum of stems and leaves is less crisped. Thus the
differences between *I. nephrocarpa* and the present plant must have a genetic basis,
and, with at least 4 clear characters involved, there seem to be ample grounds for
considering the difference to be specific.

C.3.p.4. **I. nephrocarpa** *Balf. f.*, Proc. R. Soc. Edin. **11**, 509 (1882) et Trans.
R. Soc. Edin. **31**, 73 (1888) tab. 18/A pro maj. parte.

Oman, Dhufar Mts., *Bent* 22, Yisub, 55°25′E, 17°48′N., *Thesiger* s.n. BM;
Socotra, *Schweinfurth* 237 syntype, *Balfour* 85 K lectotype BM isotype, *Popov*
SO/189 BM, EA, SO/148 BM.

C.3.p.5. **I. diphylla** *Vent.*, Choix. t. 30 (1803); L.T.A. No. 52.
Mauritania, *Roberty* 16906 K, G; Senegal (probable source of Ventenat's
unspecified type material); Fr. Sudan; Togo, *Dahomey* fide Tisserant; Fr. Niger
Col., *Hagerup* 512 BM; Nigeria, northern prov.; Fr. Chad. Terr.; A.-E. Sudan.

C.3.p.6. **I. flavicans** *Bak.*, F.T.A. **2**, 73 (1871); N. E. Brown, Kew Bull. **1909**,
102; L.T.A. No. 53.

Port. E. Afr.; S. Rhodesia; N. Rhodesia; Angola; Trop. Bechuanaland,
Baines s.n. holotype; S.W. Africa; Extra Trop. Bechuanaland; N. Transvaal.

C.3.p.7. **I. oblongifolia** *Forsk.*, Fl. Aeg. Arab. 137 (1775); L.T.A. No. 156;
Andrews, Fl. Pl. A.-E. Sudan **2**, 213, fig. 81.

I. lotoides Lam., Dict. **3**, 247 (1789), excl. syn.

I. paucifolia Del., Fl. Aeg. 107, t. 37, fig. 2, 2 (1813); F.T.A. **2**, 88.

I. desmodioides Bak., Kew Bull. **1894**, 331; non J. Linn. Soc. **22**, 463 (1887).

Egypt (*Delile* s.n. P. type of *I. paucifolia*); Jordan; Hejaz; Yemen (*Forskal*
s.n. C holotype); Aden; Hadhramaut, *Bent* 185 holotype of *I. desmodioides*;
Oman; Bahrein; Baluchistan; Sind; India; Ceylon; Java; Mauritania,
Roberty 16856K G; Senegal; Fr. Niger Col., *Hagerup* s.n. BM; N. Nigeria;
Fr. Chad. Terr.; A.-E. Sudan; Eritrea; Br. Somaliland; Socotra; Angola.

Sonnerat s.n. P, the type of *I. lotoides* is certainly this species. It probably comes
from India and not the Cape as supposed by Lamark.

I. desmodioides Bak. 1894 non 1887 is a form of this species in which the leaves
are often 1-foliolate and the fruit is more torulose than usual. It passes so gradually
into the usual form that it is not here retained even as a variety.

C.3.p.8. **I. schimperi** *Jaub. & Spach*, Ill. Pl. Or. t. 484 (1856); L.T.A. No. 158.
α. var. **schimperi.**

I. tettensis Klotzsch, Pet. Moss. Bot. 51 (1861); L.T.A. No. 159 pro maj. parte.
Edwards & Bogdan, Import. Grassland Pl. Kenya, Fig. 32, 1.

I. schimperi var. *oxyphylla* Franch. in Revoil Pays Somali Sert., 28 (1882).

A.-E. Sudan, Kassala prov., *Bent* s.n.; Ethiopia, *Schimper* (1853) 1218 P
holotype; Br. Somaliland, *Revoil* 41 P type of var. *oxyphylla*; Uganda, U1, U3;
Kenya, K1, K4, K6–7; Tanganyika, T1–2, T5–6, T8; Port. E. Afr. (*Peters*
s.n. B† type of *I. tettensis*); Nyasaland; S. & N. Rhodesia; Transvaal; Zululand.

β. var. **baukeana** (*Vatke*) *Gillett* stat. nov.

I. baukeana Vatke, Oest. Bot. Zeit. **29,** 220 (1879).

"*I. tettensis Klotzsch*" sensu Bak., L.T.A. No. 159 p.p. non Klotzsch.

Kenya, K4 *Hildebrandt* 2787 isotype, *Bogdan* 1345, 3366, *Edwards* 1938, *Mrs. M. Hale* 85, *Scott Elliot* 6759; K7 *Napier* 992, *Graham* 1621, *Drummond & Hemsley* 3744, 4088; Tanganyika, T2, *Greenway* 7491, *Volkens* 564, *Geilinger* s.n.; T3 *Greenway* 3993, 4076, 4542, 6586, *Davies* 1113, *Peter* K460, *Holst* 1075; T6 *Musk* 48; S. Rhodesia, Victoria falls, *Rogers* 5567, Wankie, *Eyles* 1299; Bechuanaland, Nakwena Territory, *Holub* s.n.

Mr. Bogdan comparing his 3366 (var. *baukeana*) with 1463 (var. *schimperi*) states that they are certainly not the same, 3366 having an erect habit, larger flowers and different pods. However, these characters are not firmly associated, each variety varies greatly and the line between them is vague.

γ. var. **crispidula** *Gillett* var. nov.

A var. *schimperi* pilis crispidulis densis, plus minusve patentibus, haud rigide appressis, differt. Stamina *c.*5 mm. longa.

A.-E. Sudan, 31 km. N. of Kapoeta, 33°29′E, 4°59′N, black cotton soil, 28.9.1953, vern. name (Taposan) NYAMARET. *A. W. Peers* 2 KAM 35. Uganda, U1, Karamoja distr., Koputh, 6.6.1942, *I. R. Dale* U270 K holo. EA isotype. Lokapeliethe, 1260 m., grassland on black cotton soil, subdominant shrub, *c.*30 cm. tall, spreading, 30.10.1939, *A. S. Thomas* 3124. Kangole 34°30′E, 2°28′N, 1200 m., overgrazed wet ground, abundant procumbent *c.*30 cm. tall, 22.5.1940, *A. S. Thomas* 3491 K, EA. District unknown, Merias, *c.*1200 m., in grass 60–90 cm. tall, flowers pink, June 1930, *Liebenberg* 322. Kenya, K2 W. Muruanisigar, *c.*35°E, 3°8′N, 1200 m., woody, flower mauve, July 1932, *A. M. Champion* T124.

δ. var. **parvifoliolata** *Gillett* var. nov.

A var. *schimperi* foliolis multo minoribus, ad 5 mm. longis, 3 mm. latis, differt.

Brit. Somaliland, Mordale, 42°38′E, 10°45′N, 740 m., sandy ground, flowers red, Mar. 1934, *H. V. Godding* 211 holotype.

In general *I. schimperi* is most variable. Some of its forms are very difficult to separate from *I. oblongifolia*.

C.3.p.9. **I. semitrijuga** *Forsk.*, Fl. Aeg. Arab. 137 (1775); L.T.A. No. 178; non sensu Bak., F.T.A. **2,** 93 et Fl. Br. Ind. **2,** 98.

I. somalensis Vatke, Oest. Bot. Zeit. **28,** 201 (1878).

Yemen, *Forskal* s.n. C holotype; Aden; Hadhramaut; Oman, Dhufar; Eritrea, *Salt* s.n. pp. BM; Br. Somaliland (*Hildebrandt* 836A B± holotype of *I. somalensis*) *Gillett* 4751, etc.

I. semitrijuga is easily confused with several other argenteous desert species especially when they are severely cut back under very arid conditions. Var. *tetrasperma* DC., Prod. (1825) is *I. argentea* Burm.f.; Var. *macrocarpa* Vatke, Oest. Bot. Zeit. **29,** 221 (1879), whose type, *Hildebrandt* 2399 has probably been destroyed, may be the same.

C.3.p. 10. **I. daleoides** *Benth. ex Harv.*, Fl. Cap. **2,** 200 (1862); L.T.A. No. 155.

α. var. **daleoides.**

I. dodecaphylla Fic. & Hiern, Trans. Linn. Soc. ser. 2, **2,** 18 (1881); L.T.A. No. 151.

I. daleoides Benth. ex Harv. var. *dammarensis* Bak.f., J. Bot. **41,** 327 (1903).

S. Rhodesia, Bulawayo, *Gardner* 57; N. Rhodesia, S. Prov., *Angus* 1067, *Gordon Read* 3; Angola (*Serpa Pinto* 7 type of *I. dodecaphylla*) *Welwitsch* 2060 BM; Trop. Bechuanaland; S.W. Africa, *Een* s.n. BM. type of var. *dammarensis*: Extra Trop. Bechuanaland; Transvaal, *Burke* 172 syntype.

β. var. **gossweileri** Bak.f., L.T.A. 138 (1926).

I. hololeuca Benth. ex Harv. var. *angolensis* Bak.f., L.T.A. 137, No. 153 (1926).

ANGOLA, Benguela, *Gossweiler* 4944 holotype, *Pearson* 2354 holotype of *I. hololeuca* var. *angolensis*.

These two specimens are not identical and their treatment here as forms of one variety of *I. daleoides* may prove to be wrong when a great deal more material of this difficult portion of the genus has been obtained from Angola.

C.3.p.11. **I. maritima** *Bak.*, F.T.A. **2,** 89 (1871); L.T.A. No. 147.

ANGOLA, *Welwitsch* 2061b isotype, etc.; S.W. AFRICA, N. of Rooivaal near Orange River, *Pillans* 6423; N.W. CAPE COL., Viols drift, in sandy bed of Khusies river, *Pillans* 6381.

C.3.p.12. **I. volkensii** *Taub.* in Engl. Pflanzenwelt O. Afr. C. 210 (1895); L.T.A. No. 157.

I. subhirtella Chiov. in Cufod. Miss Biol. Boran. 77, 5. 15. (1939).

I. boranensis Chiov. ibid. 69 t. 10 (1939).

A.-E. SUDAN, South East, *Peers* KO 17; ETHIOPIA, South, *Cufodontis* 502 FI holotype of *I. subhirtella*, 294 FI holotype of *I. boranensis*; UGANDA, U1 Karamoja distr.; KENYA, K1, K3–4, K6–7; TANGANYIKA, T1 Musoma, T2 *Volkens* 1738 isotype, T3, T5–6.

Dowson 302 from Nairobi, included in *I. phillipsiae* by E. G. Baker in L.T.A. No. 150, seems better placed in the present species.

In the northern part of its range, besides typical plants such as *Cufodontis* 502 and *Gillett* 13027 from Dandu, others occur with a much denser more appressed indumentum which approach *I. insularis* and *I. phillipsiae*. Such are *Cufodontis* 294, *Gillett* 13810 from Furroli, *Haylett* 18 from Barsaloi.

C.3.p.13. **I. insularis** *Chiov.*, Ann. Ist Bot. Rom. **8,** 418 (1908); L.T.A. No. 146.

"*I. paucifolia Del.*" sensu Terrac., Ann. Ist Bot. Rom. **5,** 109 (1894); non Del..

I. alternans DC var. *paucijuga* Schweinf., Bull. Herb. Boiss. **4,** app. **2,** 239 (1896).

"*I. alternans DC.*" sensu Chiov., Ann. Ist Bot. Rom. **8,** 87 (1903); non DC.

A.-E. SUDAN, nr. Erkowit, *c.*37°3′E., 18°45′N., *Schweinfurth* (1868) 247 isosyntype of *I. alternans* var. *paucijuga*; ERITREA, Dahlak archipelago, *Tellini* 424 FI, 662 FI, 725 FI, 762 FI syntypes.

Schweinfurth l.c. suggests that this plant may be conspecific with *I. parvula* Del. in Caill. Voy. Meroe Cent. **4,** 328 Atlas, Vol. **2,** tab. 3, fig. 1 (1826). Delile's type *Caillaud* s.n. A.-E. Sudan near the Nile, may be lost as Richard was unable to consult it when preparing his tentamen Florae Ethiopicae, no subsequent author seem to have seen it, and the authorities in Paris, who kindly loaned many specimens for the purpose of the present study, were unable to trace it. Delile's name has also been applied to a variety of *I. spicata* (*I. hendecaphylla*) although this seems almost certainly wrong in view of the nature of the stipules. On the whole it seems best to retain Chiovenda's name for the present species and to treat *I. parvula* Del. as a *nomen ambiguum* until its type can be discovered.

C.3.p.14. **I. phillipsiae** *Bak.f.*, J. Bot. **41,** 265 (1903); L.T.A. No. 150.

I. phillipsiae var. *erecta* Chiov., Fl. Somala **1,** 140 (1929).

I. phillipsiae var. *prostrata* Chiov., ibid.

BR. SOMALILAND, *James & Thrupp* s.n. syntype, *Mrs. Lort Phillips* s.n. BM syntype, etc.; SOMALIA, *Puccioni & Stefanini* 770 FI type of var. *erecta*, 86 FI type of var. *prostrata*.

C.3.p. 15. **I. sesquijuga** *Chiov.*, Fl. Somala **1**, 139 (1929).

SOMALIA, North, *Puccioni & Stefanini* 946 FI syntype.

Chiovenda's var. *obbiadensis* is related to *I. tritoides* q.v., and not to this species.

C.3.p. 16. **I. viridiflora** *Chiov.*, Fl. Somala **1**, 140 (1929).

SOMALIA, North, *Puccioni & Stefanini* 891 FI holotype.

Species 12–16 are closely related, and it may well ultimately prove more satisfactory to treat them as forms of one polymorphic species. Much more material is however necessary before any satisfactory conclusion can be reached.

C.3.p. 17. **I. spicata** *Forsk.*, Fl. Aeg. Arab. 138 (1775)

I. hendecaphylla Jacq., Coll. Bot. **2**, 358 (1788); Ic. Pl. Rar. t. 570.

"*I. endecaphylla Jacq.*" Lam Encyc. Suppl. **3**, 147 (1813) per errore; L.T.A. No. 144; F.C.B. **5**, 149; Edwards & Bogdan Impt. Grassland pl. Kenya fig. 31; 2.

I. anceps Vahl. ex Poir. in Lam. Encyc. Suppl. **3**, 147 (1813).

I. kleinii Wight & Arn., Prod. Fl. Pen. Ind. 204 (1834).

"*I. parvula Del.*" sensu Hochst. ex A. Rich., Tent. Fl. Ab. **1**, 180 (1847); non Del.

I. onobrychoides Boiv. ex Baill., Bull. Soc. L. Par. **1**, 398 (1883).

I. parkeri Bak., J. Linn. Soc. **20**, 126 (1883).

I. pectinata Bak., ibid. 127.

I. bolusii N. E. Brown in Burtt Davy Man. Flor. Transvaal **2**, XXIV, 371 (1932).

I. neglecta N. E. Brown, ibid.

I. endecaphylla var. *angustifolia* A. Rich., Tent Fl. Ab. **1**, 182 (1847).

I. endecaphylla var. *angustata* Harv., Fl. Cap. **2**, 199 (1862).

I. endecaphylla var. *acutifolia* Chiov., Ann. Ist Bot. Rom. **8**, 88 (1903).

I. endecaphylla var. *parvula* Chiov., ibid.

I. endecaphylla var. *major* Bak.f., L.T.A. 135 (1926).

YEMEN, *Forskal* s.n. C. holotype; INDIA; CEYLON (Wight Cat. 855 type of *I. kleinii*); BURMA; SIAM; PHILLIPINES; INDONESIA; SENEGAL, fide Tisserant; PORT GUINEA; LIBERIA; FR. SUDAN, fide Chevalier; IVORY COAST, fide Chevalier; GOLD COAST (*Thonning ?* s.n. C? syntype of *I. anceps*) (possibly the type of *I. hendecaphylla* is from this region); DAHOMEY; NIGERIA; ST. THOMAS; FR. CONGO; CHARI; B. CONGO; A.-E. SUDAN; ERITREA, *Schweinfurth* (1872) 1752 isosyntype of var. *parvula* (*Pappi* 283 FI type of var. *acutifolia*); ETHIOPIA, *Schimper* (1837) 366 isosyntype of var. *angustifolia*, *Schimper* (1837) 269 K, P, isotypes of "*I. parvula*" sensu Hochst. ex A. Rich.; UGANDA, U1–4; KENYA, K1, K3–7; TANGANYIKA, T1–8; PEMBA; PORT. E. AFR.; NYASALAND; S. RHODESIA, *Mrs. Macaulay* 338 type of var. *major*; ANGOLA; MADAGASCAR (*Boivin* 1910 P type of *I. onobrychoides*) *Parker* s.n. type of *I. parkeri*, *Baron* 746 type of *I. pectinata*; MAURITIUS; RÉUNION; TRANSVAAL, *Burke* s.n. type of *I. neglecta*; NATAL, *Bolus* 7722 type of *I. bolusii*, *Guenzius* 195 type of var. *angustata*.

I. pusilla Lam., Encyc. **2**, 248 (1789) has been treated as a synonym for *I. spicata*. However, although specimens from Mauritius and Réunion at Kew labelled *I. pusilla* are certainly *I. spicata* Lamark's description does not agree with them in several important respects and the species needs further investigation. *I. spicata* is most variable. "*I. parvula* Del." sensu Hochst. ex A. Rich, is a small-leaved montane form which at first sight appears specifically distinct. However, so many intermediates link it with the more widespread forms of *I. spicata* that it is unsatisfactory to treat it even as a variety. Similar considerations apply to tall plants with narrow leaves often glabrous above such as *I. pectinata*, *I. bolusii* and var. *angustata*.

I. spicata frequently is a weed and certainly owes it wide distribution in part to man. It may well be that this has obscured a previous differentiation on a geographic basis and thus given rise to the immense confusion of forms which is now observable.

C.3.p.18. **I. oxalidea** *Welw. ex Bak.*, F.T.A. **2**, 86 (1871); L.T.A. No. 143.

I. endecaphylla Jacq. var. *radicans* Welw. ex. Bak., F.T.A. **2**, 96 (1871).

I. supralevis N. E. Brown in Burtt Davy Man. Flor. Transvaal **2**, XXV, 371 (1932).

S. RHODESIA, *Swynnerton* 371, 480 BM, *Ferrar* in Rhod. Mus. 4087 SRGH, *Williams* in SRGH 27263 SRGH, *Hayter* in SRGH 18690, *Hack* in SRGH 27495, ANGOLA, *Welwitsch* 2023 isotype, 2039 type of *I. endecaphylla* var. *radicans* 4150; TRANSVAAL, *Codd* 871, 1051, *Wilms* 287 type of *I. suprelevis* 297, 312, *Rehmann* 6218, *Moss* 16149, *Rogers* 22718.

Dalziel 580 from Nigeria cited by *Tisserant* (1931) 267 as *I. oxalidea* is *I. kerstingii*. The pods of *I. oxalidea* are distinctive, though difficult to distinguish in words from those of *I. spicata*. All three Angolan specimens cited above have the stems to a greater or less extent rooting at the nodes. This character has not been seen in S. Rhodesian and Transvaal plants. *I. oxalidea* when rooting at the nodes closely resembles *I. radicifera* but may be distinguished as follows:

I. oxalidea	*I. radicifera*
Pods slightly curved, brownish.	Pods markedly curved, often semi-circular, paler.
Leaflets glabrous above.	Leaflets strigulose above.
Filaments 3–4 mm. long.	Filaments *c.* 5 mm. long.

C.3.p. 19. **I. kerstingii** *Harms*, Eng. Bot. Jahrb. **45**, 308 (1910); L.T.A. No. 8.

TOGO, *Kersting* A 662 isosyntype; NIGERIA, *Dalziel* 19, 580, *Latilo* in FHI 23548, *Olorumfemi* in FHI 24378.

C.3.p.20. **I. circinella** *Bak.f.*, J. Bot. **44**, 314 (1906); L.T.A. No. 51; Fl. Parc Nat. Alb. **1**, 296 Tab. 27 (1948); F.C.B. **5**, 148.

I. spirocarpa Harms in Mildbr. Wiss. Ergeb. Deutsch. Zentr. Afr. Exp. **2**, 252 (1911).

B. CONGO, East (*Mildbraed* 1924 B† syntype of *I. spirocarpa*), etc.; UGANDA, U2–4; KENYA, K3 *Scott Elliot* 6892 BM holo. K isotype, K4–5; TANGANYIKA, T1 (*Stuhlmann* 1814 B† syntype of *I. spirocarpa*), etc., T4.

C.3.p.21. **I. radicifera** *Cronquist*, Bull. J. Bot. Et Brux. **22**, 223 (1952); F.C.B. **5**, 148.

B. CONGO, East, fide Cronquist; TANGANYIKA, T4 *Van Meel* 1212 BR holotype, *Mrs. Loveridge* 703, *Peter* 35195; T5 *Staples* 434 K, EA; N. RHODESIA, S. end Lake Tanganyika *Mrs. Richards* 5393.

C.3.p.22. **I. alternans** *DC.*, Prod. **2**, 229 (1825); Fl. Cap. **2**, 199; L.T.A. No. 154.

I. exigua Ecklon & Zeyher, Enum. Pl. Afr. Austr. 244 (1836) (teste Harvey).

I. arenaria E. Mey. Comm. Pl. Afr. Austr. 107 (1836) non A. Rich. (1847).

I. effusa E. Mey ibid.

I. alternans var. *macra* Bak., F.T.A. **2**, 89 (1871).

ANGOLA, *Welwitsch* 2058 type of var. *macra*; S.W. AFRICA; CENTRAL and N. CENTRAL CAPE PROV., *Burchell* 2079 type, *Drége* s.n. isotype of *I. effusa* (*Ecklon & Zeyher* 1620 type of *I. exigua*); ORANGE FREE STATE; TRANSVAAL.

C.3.p.23. **I. hololeuca** *Benth. ex Harv.*, Fl. Cap. **2**,200 (1862); L.T.A. No. 152.

"*I. alternans DC.*" sensu E. Mey, Pl. Afr. Austr. 107 (1836); non DC.
S.W. AFRICA; N.W. CAPE PROV., *Drége* s.n. isotype.

C.3.p. 24. **I. benguellensis** *Bak.*, F.T.A. **2**, 87 (1871); L.T.A. No. 145.
ANGOLA, Mossamedes, *Welwitsch* 2061 type.

C.3.p. 25. **I. auricoma** *E. Mey*, Comm. Pl. Afr. Austr. 107 (1836).

I. auricoma var. *cuneata* Bak.f., Vierteljahrschr. Nat. Ges. Zur. **49**, 184 (1904); L.T.A. No. 152.

S.W. AFR. (*Nels* 263 type of var. *cuneata*), etc.; N.W. CAPE PROV., on Orange River *Drége* s.n. isotype.

Holub s.n. from Zambesi near the Victoria falls is perhaps a form of this species.

C.3.p.26. **I. antunesiana** *Harms*, Eng. B.J. **26**, 285 (1899); L.T.A. No. 148.

"*I. endecaphylla Jacq.*" "form with large leaflets" sensu Hiern, Cat. Welw. Afr. Pl. **1**, 215 (1896) non Jacq.

I. stenophylla Guill. & Perr. var. *nyassae* Bak.f., J. Bot. **41**, 261 (1903); L.T.A. 129.

"*I. trimorphophylla Taub. a close ally*" sensu Bak.f., L.T.A. 107 non *I. trimorphophylla* Taub.

B. CONGO, Elisabethville *Rogers* 10061; TANGANYIKA, T7 *Stolz* 187, T8 *Busse* 652 EA, 713 EA; PORT. E. AFR., nr. Lake Nyassa, *Johnson* 506, Tete *Torre* 6062 LISC; NYASALAND, *Buchanan* 23 BM type of *I. stenophylla* var. *nyassae*, *Exell, Mendonça & Wild* 716A BM; S. RHODESIA, Urungwe distr., *Hopkins* in SRGH 13054, Lomagundi distr., *R. W. Jack* in SRGH 9484, 9530, 11921, 15797, Salisbury distr. *Eyles* 6915, 8773 K, SRGH, Marandellas distr. *Myres* 24, *Corby* 527 SRGH; N. RHODESIA, N. Prov., *Fanshawe* 1908; S. PROV. *Allen* 204, *White* 1919; Central Prov. *Rogers* 8339, *Angus* 890; W. Prov. *Kassner* 2214; ANGOLA (*Antunes* s.n. COI syntype) (*Melho Ramalho* s.n. COI syntype); *Welwitsch* 2014, *Gossweiler* 2363, 4141, 9300. *Faulkner* 25, 490 ($\pm$ stunted forms) *Hundt* 738 BM. This species is variable; particularly in the relative lengths of calyx and corolla. The stiffly erect habit, woody rootstock, circinnate fruits and particularly the peculiar short stiff hairs of the upper leaf surface are characteristic.

C.3.p.27. **I. conjugata** *Bak.*, F.T.A. **2**, 75 (1871); L.T.A. No. 54; F.C.B. **5**, 150.

α. var. **conjugata.**

I. dalzielii Hutch., Kew Bull. **1921**, 246 pro maj. parte; L.T.A. No. 46 p.p.

I. dalzielii var. *congolensis* Bak.f. ex De Wild., Pl. Beq. **2**, 530 (1924).

"*I. knoblecheri*" sensu Chevalier, Expl. Bot. Afr. Occ. Fr. 174 non Kotschy.

FR. GUINEA, *Chevalier* 340 P, 364 P; DAHOMEY, *Chevalier* 23913 P; 24066 P; NIGERIA, *Hill* 4, *Lely* 20 syntypes of *I. dalzielii*, etc.; B. CONGO, *Elskens* 135 BR type of *I. dalzielii* var. *congolensis*, etc.; A.-E. SUDAN, c.31°E, 5°N, *Petherick* s.n. holotype; UGANDA, U1, U3–4; KENYA, K3, K5; TANGANYIKA, T1, T4; ANGOLA, *Young* 790 BM.

β. var. **trimorphophylla** (*Taub.*) *Gillett* stat. nov.

I. trimorphophylla Taub., Eng. B.J. **23**, 182 (1896); *R. E. Fries*, Schwed. Rhod. Kong. Exp. 77 (1916); L.T.A. No. 43.

B. Congo, Urundi, Mecherenge-Ruchivoka *R. E. Fries* 1451 UPS neotype; N. Rhodesia, Mwinilunga distr., *Milne-Redhead* 2613; Angola (Malansche *Mechow* 276 B† holotype).

R. E. Fries states that he compared his 1451 with Mechow's type and that it "gut ubereinstimmt". *I. trimorphophylla* is not mentioned in the Flore du Congo Belge and is, indeed, perhaps hardly worth retaining, even as a variety.

γ. var. **occidentalis** *Gillett*, Kew Bull. **1955**, 581 (1956).

I. dalzielii Hutch., Kew Bull. **1921,** 246, F.W.T.A. **1**, 393 et L.T.A. No. 46 pro minore parte.

French Sudan; N. Nigeria; Brit. Cameroons, Gashaka distr., Mai Idoanu *Latilo & Daramola* in FHI 34470.

δ. var. **schweinfurthii** (*Taub.*) *Gillett* stat. nov.

I. schweinfurthii Taub., Eng. B.J. **23,** 181 (1896); L.T.A. No. 45.

A.-E. Sudan, Djurland, *Turner* 279, *Schweinfurth* 1774 isosyntype, and lectotype.

C.3.p.28. **I. chevalieri** *Tiss.*, Bull. Mus. Hist. Nat. Par. Ser. **2**: **3**, 163 (1931).

E. Oubangui-Chari, Ndellé, 20°20′E, 8°15′N, *Chevalier* 8132 P. holotype.

Closely related to *I. conjugata* var. *schweinfurthii*: further material may show it to be but a form of that plant. Besides the narrower leaflets the leaf rhachis seems to be shorter and the inflorescence often longer than in var. *schweinfurthii*.

Subsect. C.3.q. Simplices-reflexae subsect. nov.

Type species *I. bongensis* Kotschy & Peyr.

Centre of distribution the A.-E. Sudan.

Folia semper simplicia. Calycis lobi tubo multuplo longiores. Legumina (ubi cognita) recta, longa, reflexa, endocarpio maculato.

This subsection is closely related to *I. conjugata* Bak. in subsect. **Alternifoliolae.**

Hairs on stems, inflorescence, etc. $\pm$ crisped and spreading; leaves usually glabrous above, except for the margins; calyx shorter than stamens:

  Leaves narrowed at the base, margin and midrib forming an angle of 30° or less
conjugata var. occidentalis.

  Leaves broader at base, margin and midrib forming an angle of *c*.45° or more
1. bongensis.

Hairs on stems, inflorescence, etc., appressed; leaves strigulose above:

  Calyx from one-half to three-quarters as long as keel . . . . 2. *knoblecheri*.

  Calyx longer than corolla:

    Inflorescence not very compact at time of flowering, usually longer than the *c*.6 cm. long subtending leaf; calyx teeth whitish, shorter than the erect deciduous bracts . . . . 3. *achyranthoides*.

    Inflorescence very compact at time of flowering, about half as long as the up to 9 cm. long subtending leaf; calyx teeth blackish, longer than the subpersistent, later reflexed, bracts . . . . 4. *hundtii*.

C.3.q.1. **I. bongensis** *Kotschy & Peyr.*, Pl. Tinn. **8,** tab. 4 (1867); L.T.A. No. 40; Tisserant, Bull. Mus. Hist. Nat. Par. 2nd ser. **2**, 681 (1930), F.C.B. **5**, 131.

Oubangui, South-West, *Mildbraed* 9453, *Tessman* 2278, *Lenfant* 1091 P; Chari, East, *Chevalier* 7424 (approaching I. *I. conjugata* var. *occidentalis*); B. Congo,

122

North, fide Cronquist; A.-E. Sudan (1863 *De Heuglin* s.n. W holotype) Jurland *Schweinfurth* 1899, 1967; Wau district *Mackintosh* 5, 30, 32; Equatorial Prov. *Myers* 7001, etc.

C.3.q.2. **I. knoblecheri** *Kotschy*, Sitzb. Kais. Ak. Wien. Math. Nat. **1.** Abt. **1**, 363 (1865); Schweinfurth, Reliq. Kotsch. 19, t. 15; L.T.A. No. 41.

A.-E. Sudan, Gondokoro, 5°N, (*Knoblecher* s.n. W. holotype), Sue river E. of Zbue Satta hills, *c.*27°50′E, 6°10′N, occasional in *Combretum*-grass woodland, 11.7.1937, *Myers* 7097.

C.3.q.3. **I. achyranthoides** *Taub.*, Eng. B.J. **23**, 180 (1896); L.T.A. No. 33; F.C.B. **5**, 131.

I. wauensis Schweinf. ex Cronquist, Bull. J. Bot. Et. Brux. **22**, 226 (1952).

B. Congo, North, fide Cronquist; A.-E. Sudan, Djurland on Wau river, *Schweinfurth* 1624 isotype of both names.

C.3.q.4. **I. hundtii** *Rossberg*, Fed. Rep. **38**, 106 (1935).

Angola, Benguella, Ganda-Caconda, 1700 m. *Hundt* 559 BM isotype.

The fruits are unknown.

Subgenus D. Indigastrum (Jaub. & Spach) Gillett stat. nov.

Indigastrum Jaub & Spach pro gen., Ill. Pl. Or. **5**, 101, t. 492, 493 (1856).

Parviflorae Rydberg, N. Amer. Flor. **24**, 140 (1923).

"group" *Indigastrum* (J. & S.) Bak.f. L.T.A. 161 (1926); Cronquist F.C.B. **5**, 124 p.p.

Type species *I. parviflora* Heyne ex Wight & Arn. (*Indigastrum deflexum*).

Centre of distribution, Southern Africa.

Leaflets 1–17, the lateral ones usually opposite, the rhachis usually prolonged beyond the last pair. Indumentum always appressed, inflorescence an axillary raceme, dense in the bud but usually opening out as the fruits mature. Fruiting pedicel reflexed, bracts caducous. Corolla glabrous, the keel rostrate, somewhat gibbous but not spurred at the sides. All stamens fertile, the anthers usually without basal scales (these are present in No. 2). Pod somewhat flattened, linear, many-seeded, usually ± reflexed and pointing in the same direction as the upper part of the pedicel, endocarp not spotted. Seeds many, close together.

Dr. Cronquist, although he does not deal with any species hitherto named **Microcharis,** includes all the Belgian Congo species which are here considered to belong to **Microcharis** (*I. medicaginea, I. welwitschii, I. asparagoides, I, tisserantii, I. praetermissa*) in his **Indigastrum.** There is, in fact, much to be said for this point of view and the two groups are certainly closely related. If they are kept separate, as here, **Indigastrum** must be considered as a bridge between **Microcharis** and the rest of the genus.

Inflorescence pedunculate, the peduncle longer than a leaflet; leaflets opposite:
  Terminal leaflet usually longer than the combined petiole and rhachis:
    Leaflets 3, thinly silky above . . . . 1. *crotalarioides*.
    Leaflets 3–7, glabrous above or nearly so, usually 6–10 times as long as wide; lower calyx lobes twice as long as tube, corolla 5–7 mm. long; the two anthers next to the median ventral anther having basal appendages twice as long as themselves; style at an angle of less than 90° with the ovary; perennial . . . . 2. *fastigiata*.

123

Terminal leaflet usually shorter than the combined petiole and rhachis, leaflets rarely more than 6 times as long as wide, calyx lobes rarely twice as long as calyx tube; basal appendages much shorter than anthers, or absent; style at an angle of 90° with ovary:

Corolla 7–11 mm. long; leaflets 5–7, rarely 3 or 9, usually densely appressed strigulose above; pods straight; perennial . . . . 3. *burkeana*.

Corolla up to 6 mm. long; pods usually slightly upturned near the tip; usually annual . . . . 4. *costata*.

Upper surface of leaflets sparsely appressed-strigulose:

Leaflets 11–15 . . . . α. ssp. *costata*.

Leaflets 5–11 . . . . β. ssp. *theuschii*.

Upper surface of leaflets glabrous:

Leaflets 9–15 . . . . γ. ssp. *goniodes*.

Leaflets 5–9, usually 7 . . . . δ. ssp. *macra*.

Inflorescence ± sessile, shorter than the subtending leaf, the peduncle, if present, shorter than a leaflet; corolla under 7 mm. long:

Leaflets 3–13, opposite, usually more than 3 times as long as broad, if wider, then glabrescent above; pod usually over 20 mm. long; annual . . . 5. *parviflora*.

Hairs on young fruit straight, appressed:

Leaves 3–6-jugate, rarely 2-jugate, usually glabrous above, or nearly so; pods strongly reflexed . . . . α. var. *parviflora*.

Leaves 1–3-jugate, appressed-strigose above, pods erect, spreading or reflexed . . . . β. var. *occidentalis*.

Hairs on young fruits crisped, spreading at the tips; leaflets 3–4-jugate, appressed-strigulose above . . . . γ. var. *crispidula*.

Leaflets 1–6, usually alternate, densely appressed-strigulose above; pod not over 20 mm. long:

Annual; leaflets 1–3, the terminal much larger than the others, orbicular-obovate, 1–2 times as long as broad, up to 20 mm. long and 15 mm. wide; style about as long as ovary . . . . 6. *argyroides*.

Perennial; leaflets 3–6, rarely 2, the terminal not much larger than the others, obovate-oblanceolate, 2–3 times as long as broad, up to 13 mm. long; style *c*. twice as long as ovary . . . . 7. *argyraea*.

D.1. I. crotalarioides (*Klotzsch*) *Bak.*, F.T.A. **2,** 85 (1871); L.T.A. No. 160.

Tephrosia crotalarioides Klotzsch, Pet. Moss. Bot., 45 (1861).

PORT. E. AFR., Sena, in sandy fields (*Peters* s.n. B† holotype) (probably Sena island in Mozambique harbour 15°5′S, possibly Sena point 15°29′S).

It is desirable that a search should be made for this plant in the type locality. If its corolla is indeed glabrous, as described, it is probably a member of subgen. **Indigastrum,** and may be but a form of *I. fastigiata*. This latter species has however not been collected from Portuguese East Africa. The species has been said to be like *I. trita* and *I. spiniflora* but there are serious discrepancies between its description and both these species.

D.2. I. fastigiata *E. Mey.*, Comm. Pl. Afr. Austr., 102 (1836); Harvey, Fl. Cap. **2,** 183.

I. rostrata Conrath, Kew Bull. **1908,** 223, non Bolus, J. Bot. **34,** 23 (1896).

I. fastigiata var. *angustata* Harv., Fl. Cap. **2,** 183 (1862).

TRANSVAAL, *Conrath* 1196 holotype of *I. rostrata* Conrath, etc.; NATAL; ORANGE
FREE STATE; BASUTOLAND; E. CAPE PROV., *Drége* s.n. isotype, *Hutton* s.n. syntype of
var. *angustata*.

D.3. **I. burkeana** *Benth. ex Harv.*, Fl. Cap. **2**, 197 (1862).

I. affinis Harv., Fl. Cap **2**, 184 (1862) non De Wild., Bull., J. Bot. Et. Brux **8**,
144 (1923).

I. goniodes var. *damarensis* Bak.f., L.T.A., 163 (1926).

TROP. BECHUANALAND, *J. McCabe* s.n. holotype of *I. affinis* Harv.; S.W. AFRICA
Een s.n. BM holotype of *I. goniodes* var. *damarensis, Chapman & Baines* s.n.,
Pearson 4648; EXTRA TROP. BECHUANALAND, *Holub* s.n., *Mogg* 8043; TRANSVAAL,
Burke & Zeyher 476 type, etc.

McCabe s.n. and *Pearson* 4648 have leaflets which are glabrous above and in this
respect approach *I. costata* ssp. *goniodes* and *macra*.

D.4. **I. costata** *Guill. & Perr.*, Fl. Seneg. 187 (1832); L.T.A. No. 247.

α. ssp. **costata**.

SENEGAL, *Perrottet* s.n. BM iso. (P holotype), *Huard* in Heudelot 508; *Berhaut*
569 P; N.E. NIGERIA, *Mrs. Noble* 28; NIGER-CHAD, *Gaillard* in Mission *Tilho* s.n.
P; A.-E. SUDAN, Darfur Prov., *Mrs. Macintosh* 127 *Lynes* 519.

I. costata is described as having a short spur at the side of the carina. However
in the specimen which I dissected (*Huard in Heudelot* 508) no spur was found but
only the bagginess normal in subgenera **Indigastrum** and **Microcharis**. The
distinction between sspp. *costata* and *theuschii* is not great and may well prove
untenable when further material is available.

β. ssp. **theuschii** (*O. Hffm.*) *Gillett* stat. nov.

I. theuschii O. Hoffm., Linnaea **43**, 126 (1880–82), L.T.A. No. 243.

I. heptaphylla Hiern, Cat. Welw. Afr. Pl. **1**, 209 (1896).

ANGOLA, Pungo Andongo, *Theusch. & Mechow* 128 (B† holo.) BM photo.
Welwitsch 4143 BM, 4144 BM, 4145, 4146, 4147 BM 4148 BM, isosyntypes of
I. heptaphylla, Mrs. Faulkner A275, etc.

γ. ssp. **goniodes** (*Hochst. ex Bak.*) *Gillett* stat. nov.

I. goniodes Hochst. ex Bak., F.T.A. **2**, 85 (1871); L.T.A. No. 246.

Indigastrum macrostachyum Jaub. & Spach, Ill. Pl. Or. t. 493 (1857) non *Indigofera macrostachya* Vent., Jard. Malm. t. 44 (1803).

ERITREA, *Pappi* 354; ETHIOPIA, Dschadscha, 28.9,1854, *Schimper* 2242 holotype,
Dschadscha Aug. 1853, (*Schimper* P holotype of *Indigastrum macrostachyum*);
UGANDA, U1, Chua distr., LIEBENBERG 227; KENYA, K3, K4, K6; TANGANYIKA,
T2, T3.

δ. ssp. **macra** (*E. Mey*) *Gillett* stat. nov.

I. macra E. Mey., Comm. Pl. Afr. Austr. 105 (1836); Fl. Cap. **2**, 95.

I. gonioides var. *rhodesica* Bak.f., J. Bot. **58**, 76 (1920).

PORT. E. AFR., South of 23°S, *Gomes e Sousa* 158, etc.; S. RHODESIA, *Rogers*
13666 BM and *Gardner* 62 syntypes of *I. gonioides* var. *rhodesica*, etc.; N. RHODESIA,
North-West, Kalomo, *Rogers* 7717 BM; S.W. AFRICA; TRANSVAAL; NATAL;
E. CAPE PROV., *Drége* s.n. isotype.

Many of the southern Rhodesian specimens are intermediate between ssp.
macra and ssp. *goniodes*, having leaves occasionally 5-jugate. At first sight ssp.
macra seems very distinct from ssp. *costata* but each passes so gradually into ssp.
goniodes that it is impossible to keep them specifically distinct.

D.5. **I. parviflora** *Heyne ex Wight & Arn.*, Prod. Fl. Pen. Ind. Or. 201 (1834); L.T.A. No. 245; F.C.B. **5**, 170.

α. var. **parviflora.**

I. machaerocarpa Fenzl. nomen.

I. deflexa Hochst. ex A. Rich., Tent. Fl. Ab. **1,** 178 (1847).

Indigastrum deflexum (Hochst. ex A. Rich.) Jaub. & Spach, Ill. Pl. Or. t. 492 (1857).

Indigofera barcensis Chiov., Att. Soc. Nat. Modena **63,** 36 (1932).

Peninsular India, *Heyne* in Wall. Cat. 5457 type, etc.; N.W. Australia; B. Congo, East; A.-E. Sudan, *Kotschy* 118, *I. machaerocarpa, Kotschy* 14 isosyntype of *I. deflexa*, etc.; Eritrea, *De Benedictis* 200 FI holotype of *I. barcensis*; Ethiopia, *Schimper* (1840) 1467 isosyntype of *I. deflexa*; Uganda, U2; Kenya, K4–5; Tanganyika, T1–2; S. Rhodesia; S.W. Africa; Transvaal; N. Central Cape Prov.; Orange Free State; casual in Alabama fide Rydberg.

β. var. **occidentalis** *Gillett*, Kew Bull. **1955,** 582 (1956).

"*I. linearis*" sensu Guill. et Perr., Fl. Seneg. 184 (1832); L.T.A. No. 244: non D.C., Prod. **2,** 228 (1825).

"*I. parviflora* Heyne" sensu Hutch. & Dalz., F.W.T.A. **1,** 391; Berhaut, Fl. Seneg. 53: non Heyne ex Wight & Arn.

Cape Verde Isles; Senegal, *Leprieur* s.n. BM iso (P holotype); N. Rhodesia; Angola; N. Central Cape.

The parallel variation in *I. costata* and *I. parviflora* is remarkable. In both species the forms with leaflets strigulose above are found in West Africa and in Angola and adjacent areas, forms with leaflets glabrous above in the east.

γ. var. **crispidula** *Gillett* var. nov.

A varietatibus aliis duobus planta tota robustiore et pilis leguminis junioris crispidulis, apicibus patentibus differt. Foliola 5–9, ad 30 mm. longa, 9 mm. lata, utrinque strigulosa. Pedicelli fructiferi reflexi.

S. Rhodesia, Matobo distr., 1350 m., 11.3.1947, *O. West* 2252 K holo. SRGH isotype. Insiza, *c.*29°10′E, 19°45′S, Feb. 1909, *Allen* 733.

D.6. **I. argyroides** *E. Mey.*, Comm. Pl. Afr. Austr. 106 (1836); Fl. Cap. **2,** 191.

I. saxicola Engl., Engl. B.J. **10,** 28 (1888), non F. Muell (1864).

I. engleri Bak.f., J. Bot. **41,** 194 (1903); L.T.A. No. 31.

S.W. Africa, Lower Swachaub river, 250 m., *Marloth* 1209 (B†), BM sketch, type of *I. saxicola & I. engleri* (presumably the Swakop river *c.*14°50′E, 22°42′S). Kl Karas, 14.11.1923, *Dinter* 5096, Little Karas Mts. nr. Holoog *Pearson* 9754, 12 km. W. of Sandverhaar *Pearson* 4599 between Assenkyer and Ai Ais, 360 m., 29.12.1949, *Mrs. Macdonald* 121 BM; N.W. Cape Prov., West of Pella, *Pearson* 3561, Banks of Orange (Garip) river 70 m. *Drége* s.n. isotype, *Wyley* 52.

D.7. **I. argyraea** *E. & Z.*, Enum. Pl. Afr. Austr. 239 (1836); Fl. Cap. **2,** 201.

I. collina E. & Z., ibid. (fide Harvey).

"*I. burchellii D.C.*" sensu E. Mey., Comm. Pl. Afr. Austr. 106, non DC.

S.W. Africa, little Karas, *Dinter* 5079; Orange Free State; Basutoland; W. Natal; N.W. Cape; N. Central Cape; S. Central Cape, *Ecklon & Zeyher* 1595 isotype.

Subgenus E. **Microcharis** (*Benth.*) *Gillett.*

Microcharis (as genus) Benth. in Benth. & Hook.f. Gen. Pl. 501 (1865); L.T.A. 256.

Subgen. *Indigastrum* sensu Cronquist, F.C.B. **5**, 124 pro parte.

Type species, *I. hutchinsoniana* Gillett (*M. tenella* Benth.).

Centre of distribution, Central Africa.

Leaves simple or 1–9-foliolate, lateral leaflets, where present, opposite. Inflorescence a lax raceme, bracts persistent, usually shorter than the pedicels which most often are long for the genus. Fruiting pedicel usually $\pm$ spreading (neither erect nor deflexed) the fruits often at right angles to the pedicel. Corolla glabrous, keel nearly always rostrate, somewhat gibbous at the sides but not spurred. Stamens not above 4 mm. long; Vexillar stamen sterile (except in sp.l.); anthers, at least those of the shorter stamens, with hyaline, more or less entire, or forked, scales at the base (prolongations of the connectives). Style short, thick, widened and laterally compressed at the base; stigma large for the genus. Fruit linear, somewhat flattened, 3–30-seeded, the seeds flattened crowded close together with thin septa between them. Endocarp not spotted.

Until now **Microcharis** has been separated from **Indigofera** as having basifixed instead of medifixed hairs and anthers without an apical appendage. The first of these characters is wholly valueless, since even in the type material of *M. tenella* medifixed hairs are abundant on the undersides of the leaves. It is true that most of the hairs on the upper sides of the leaves appear simple because one arm is much reduced or obsolete. But similar hairs are found in many species of **Indigofera** s.str., particularly in **Latestipuatae.** The apical process is indeed absent from the anthers of *M. tenella* but it is present in at least some of the anthers of all the other species here included in **Microcharis,** so far as it has been possible to examine them. It is therefore not surprising that several botanists, including Engler, E. G. Baker and Merxmueller have suggested that the relationship between **Indigofera** and **Microcharis** requires examination. There can be no doubted that **Michrocharis** as here treated is a very natural group of species. It shows indeed two important characters which have hitherto been overlooked, the sterile vexillar stamen (in all species except *I. spathulata*) and the hyaline scale present, in every species, at the bases of all or some of the anthers, which however are also found outside **Microcharis** in *I. fastigiata* (Subgen. **Indigastrum**). The former character has been overlooked because, as some anthers usually break off in opening the flower, it is necessary to count them to make sure that 9 only are present: for this reason artists, unless warned, usually "replace" one of the loose anthers on the vexillar filament. The basal hyaline appendages are also difficult to observe before the anthers are removed, and afterwards are readily mistaken for a detached piece of filament.

On the whole it seems debatable whether **Microcharis** is better placed as a subgenus of **Indigofera** or as a related but distinct genus. The former course has been adopted here (1) because subgenus **Indigastrum** provides a connecting link between **Microcharis** and **Indigofera** s. str. and (2) because it is in practice more convenient (involving fewer name changes).

Leaves all simple:
  Petiole much shorter than stipules, not separated from lamina by a distinct joint; seeds often more than 15:
    Usually annuals; indumentum relatively sparse, leaves not appearing silvery:
      Leaves almost or quite glabrous; fruit glabrous or almost so; stipules not cordate:
        Leaves not cylindrical:

I

Corolla *c.*3 mm. long, leaves spathulate, up to 15 mm. wide . . . .
. 1. *spathulata.*

Corolla over 6 mm. long, leaves lanceolate, *c.*8 mm. wide . . . .
. 2. *brevistaminea.*

Leaves cylindrical, up to 10 cm. long . . . . . 11.β. *phyllogramme* var.
aphylla.

Leaves not glabrous, not spathulate:

Fruit not glabrous; stipules not broadly and cordately lobed:

Calyx longer than androecium; leaves up to 8 mm. broad. 3. *longicalyx*
Calyx shorter than androecium:

Hairs on stems and rhachis of inflorescence ± spreading; leaves up
to 9 mm. broad . . . . 4. *richardsiae.*

Hairs on stems and rhachis of inflorescence appressed; leaves
2–2·5 mm. broad . . . . 5. *buchneri.*

Fruit glabrous; calyx less than half as long as stamens; stipules broadly
cordately lobed at one side; leaves often over 15 mm. broad . . . . .
6. *lobata.*

Perennials; indumentum usually relatively dense, so that the leaves appear
silvery, at least when young, if indumentum sparse, then stipules broad
and cordately lobed; fruit not glabrous:

Leaves ± narrowly elliptical, less than 3 times as long as wide; stipules ±
cordate at the base, less than 5 mm. long . . . . 7. *microcharoides.*

Leaves linear lanceolate, more than 3 times as long as wide; stipules linear
lanceolate, 6–11 mm. long . . . . 8. *stipulosa.* .

Petiole separated from lamina by a distinct joint, much longer than stipules
except in No. 9; seeds 15 or fewer:

Stems not glabrescent:

Hairs on stems ± spreading; petiole shorter, or hardly longer than stipules;
racemes usually longer than subtending leaf . . . . . 9. *hutchinsoniana.*

Hairs on stems dense, closely appressed, petiole much longer than stipules;
racemes usually shorter than subtending leaf . . . .
10. *disjuncta* var. *disjuncta.*

Stems glabrescent; racemes longer than the leaves:

Seeds 10–15, stipules up to 1 mm. long . . . . 11. *phyllogramme* var.
phyllogramme.

Seeds 3–9, stipules 1·5-4 mm. long . . . . 12. *pseudoindigofera.*

Leaves, or at least the upper ones, compound:

Upper leaves digitately 3-foliolate:

Leaflets linear-lanceolate, many times as long as wide; stems glabrous or with
a few appressed hairs:

Stipules 1·5–4 mm. long, rarely half as long as petiole, which is 4–11 mm.
long; seeds 3–9 . . . . 12. *pseudoindigofera.*

Stipules up to 1 mm. long, much shorter than petiole, seeds 10–20 . . . . .
13. *annua.*

Leaflets obovate-lanceolate or oblanceolate, not more than 3–4 times as long as
wide, or, if longer, then petiole less than twice as long as stipules; stem
argenteous or with spreading hairs:

Fruit contorted, torulose, with elongated seeds, plant densely argenteous
14. *contorta.*

Fruit straight or curved, seeds $\pm$ flattened and square:

Perennial or, if flowering in first year, densely argenteous:

Plant silvery, hairs dense thick appressed; leaflets suborbicular, cuneate-obovate or cuneate-oblanceolate, usually less than twice as long as broad:

Leaves mostly simple; fruiting pedicel under 2 mm.; seeds fewer than 14; inflorescence shorter than the leaves; leaflets under 6 mm. long . . . . 10. *disjuncta* var. β *fallax*.

Leaves mostly 3-foliolate; fruiting pedicel *c.*3 mm.; seeds more than 14; inflorescence longer than the leaves . . . . . 15. *tritoides*.

Leaflets over 5 mm.; inflorescence over 2 cm. long . . . . α. var. *tritoides*.

Leaflets under 5 mm.; inflorescence under 2 cm. long . . . . β. var. *obbiadensis*.

Plant green, hairs relatively sparse and spreading; leaflets oblanceolate, usually more than twice as long as broad . . . . . . 16. *cufodontii*.

Delicate annual, hairs sparse, somewhat spreading, leaflets lanceolate . . . 17. *praetermissa*.

Upper leaves pinnately 3-foliolate, or pinnate:

Plant not glabrous; leaflets 1–5, linear-lanceolate to obovate:

Indumentum sparse, often appressed; fruit (where known) rarely reaching 3 cm. in length; mature inflorescences under 12 cm. long:

Leaflets 3–5, more than twice as long as wide; hairs long, appressed or spreading . . . . 18. *welwitschii*.

Leaflets narrow linear-lanceolate tapering gradually at both ends . . . . α. var. *welwitschii*.

Leaflets oblong-lanceolate, up to 5 times as long as wide, narrowing $\pm$ abruptly at both ends . . . . β. var. *remotiflora*.

Leaflets 3, less than twice as long as wide; hairs short, closely appressed:

Inflorescence up to 4 cm. long; pedicels up to 10 mm. long; calyx lobes about as long as tube . . . . 19. *medicaginea*.

Inflorescence up to 2 cm. long; pedicels up to 2 mm. long; calyx lobes about twice as long as tube . . . . 20. *butayei*.

Indumentum copious, spreading; fruits over 3 cm. long; mature inflorescence up to 30 cm. or more long; leaflets 1–5 . . . . 21. *wajirensis*.

Plant glabrous, or almost so, leaflets acicular, 2–6 jugate:

Pod straight, 12–25 mm. long; seeds 6–14, *c.*1 mm. long. . . 22. *asparagoides*.

Pod curved, 8–10 mm. long; seeds 3–5, 1·5–3 mm. long . . . 23. *tisserantii*.

E.1. **I. spathulata** *Gillett* sp. nov.

Species propter folia fere glabra, spathulata, bene distincta.

Herba, ut videtur annua, vel forsan basi sublignosa interdum perennante, ramis patentibus viridibus glabris subcostatis ad 7 cm. longis. *Folia* fere glabra, pilis raris, biramosis, appressis, *c.*0·5 mm. longis, ad paginam inferiorem solam conspersa, spathulata, ad 3 cm. longa et 15 mm. lata, apice rotundata, petiolo 1·5 mm. longo, stipulis subulatis, glabris, 3–5 mm. longis. *Racemi* 3–6-flori, fere glabri, pilis perpaucis appressis exceptis, ad 6 cm. longi, pedunculo ad 3·5 cm. longo incluso, pedicellis glabris, ad 3 mm. longis, post anthesin patentibus, bracteis anguste lanceolatis, *c.*1 mm. longis, persistantibus. *Calyx* fere glaber, tubo 1 mm. longo, dentibus deltoideis 0·6 mm. longis. *Vexillum* glabrum, apice retusum, oblongum, 2 mm. latum, 3 mm. longum, unguiculo lato 0·5 longo incluso. *Alae* glabrae, oblongae, basin auriculatae, 1 mm. latae, 2·9 mm. longae, unguiculo 0·3 mm. longo incluso. *Carina* glabra, rostrata, haud calcarata, 3 mm. longa, unguiculo 1 mm. longo incluso. *Stamina* omnia fertilia, *c*.2 mm. longa, antheris ellipticis, apice apiculatis, basin squamo angusto hyalino, apice fimbriato, appendiculatis, in tota *c*.0·5 mm. longis; squamae basales antherum filamentis brevioribus quam alias aliquot longiores. *Ovarium* glabrum, 1·2 mm. longum, stylo valde complanato, *c*.0·8 mm. longo, ad angulam 30° inclinato, stigmate discoideo 0·25 mm. diametro. *Legumen* immaturum glabrum, rectum, ± complanatum, 22 mm. longum, 0·9 mm. latum, 19-spermum.

N. Rhodesia, Mweru-Luapula distr., near Mbereshi, *c*.28°50′E, 9°45′S, 1170 m., small plant in short grass, flowers yellow fairly common, Feb. 1932, *Miss E. G. Walter* 7 holotype. Near Abercorn, pans, 1500 m., in gritty sandy soil among flat ironstone rocks, flowers deep pink, 18,2,1955, *Mrs. Richards* 4555. 1470 m., in gritty soil, annual, flowers rose pink, 21.2.1955, *Mrs. Richards* 4604. Lunzua Falls, 1500 m., in yellow sandy soil along track, flowers bright pink, 7.4.1955, *Mrs. Richards* 5353. 1500 m., in semi-swamp, *Exell, Mendonça & Wild* 1252 BM, 1263 BM.

This is the only species in Subgen. *Microcharis* where the vexillar stamen has been found to be fertile.

E.2. **I brevistaminea** *Gillett* sp. nov.

Species propter folia lanceolata fere glabra, petiolo haud articulato, stamina que quam carina longe rostrata multo breviora bene distincta.

Herba forsan annua, ramis costatis fere glabris, pilis medifixis appressis *c*.0·5 mm. longis sparsissime obsitis. *Folia* simplicia oblanceolata, apice acuta, basin attenuata, fere glabra, *c*.3 cm. longa, 8 mm. lata, nervis superne impressis, inferne prominentibus, petiolo *c*.1 mm. longo haud articulato, stipulis anguste deltoideoacuminatis, ad 4 mm. longis. *Racemi* fere glabri, laxe multiflori, *c*.14 cm. longi, pedunculo 4–5 cm. longo incluso, bracteis lanceolatis, 1 mm. longis, persistentibus, pedicellis e rhachide rectangulare patentibus, post anthesin ad 4 mm. longis. *Calyx* sparse appresseque strigulosus, tubo 1 mm. longo, lobis deltoideis aequilongis, *c*.1·5 mm. longis, duo superioribus quam aliis paulo latioribus. *Vexillum* glabrum, cuneato obovatum, apice emarginatum, 0·5 mm. unguiculatum, 6.5 mm. longum, 6 mm. latum. *Alae* glabrae, 8 mm. longae, unguiculo 1 mm. longo incluso, lamina cuneata, apice rotundata, basin 1·2 mm. lata, apicem versus 4·5 mm. lata. *Carina* glabra, haud calcarata, 2 mm. unguiculata, *c*.2 mm. tenuiter rostrata, in tota 6·5 mm. longa. *Stamen* liberum dorsale sine anthera; stamina fertilia, filamentis 3·2–3·6 mm. longis, post anthesin valde curvatis, antheris pallidis oblongis, basin et ad apicem squamis hyalinis praeditis, squamis basalibus paulo fimbriatis, eis quatuor antherarum ad filamenta breviora quam quinque aliis longioribus, anthera ipsa *c*.0·4–0·5 mm. longa, squamo apicale 0·1–0·2 mm. longo, basale 0·2–0·4 mm. longo. *Ovarium* glabrum 2·2 mm. longum *c*.22-ovulatum, ovulis biserriatis, stylo quam ovarium latiore, ab eo ad angulum 150° producto, valde complanato, 1·2 mm. longo, stigmate discoideo 0·7 mm. diametro. *Legumen* valde immaturum glabrum angustum, basin curvatum.

B. Congo, N.E. of Lake Moero, Niembe river, under trees, 27.5.1908, *Kassner* 3009 K holo. BM isotype.

The Brussels sheet of this number may not be the same as it was identified by De Wildeman as *I. welwitschii* and by Dr. Cronquist as *I. welwitschii* var. *remotiflora*.

E3. **I. longicalyx** *Gillett*, Kew Bull. **1955,** 583 (1956).

I. lateritia Chev., Expl. Bot. Afr. Acc. **1,** 174, non Willd.

"*Microcharis tenella Benth.*" sensu Bak.f., L.T.A. p. 256 p.p.; Hutch & Dalz., F.W.T.A. **1,** 388 p.p.; non Benth.

FR. GUINEA; SIERRA LEONE; N. NIGERIA; OUBANGUI, *Bozoum* distr. *Tisserant* 2993.

E.4. I. richardsiae *Gillett*, Kew Bull. **1955**, 583 (1956).

Microcharis angolensis Bak., F.T.A. **2**, 132 probably, non *Indigofera angolensis* D. Dietr.

TANGANYIKA, T1; PORT. E. AFR.; NYASALAND; N. RHODESIA; ANGOLA.

E.5. I. buchneri *Taub.*, Eng. B.J. **23**, 180 (1896); L.T.A. No. 34.

ANGOLA, Malange (Malandje) edge of sugar cane field, March, *Buchner* 644 (B† holotype) BM sketch.

It may be that this will prove to be conspecific with *Microcharis angolensis* which comes from the same area. However the published description shows the differences listed below. While such differences might well occur within the limits of a single species, it seems preferable to keep the two plants distinct until much more material is available.

<table>
<tr><td>M. angolensis
(Welwitsch 2004)</td><td>I. buchneri e descriptione</td></tr>
<tr><td>1. Lower leaves up to 7 mm. wide.</td><td>"folia 2–2·5 mm. lata".</td></tr>
<tr><td>2. Plant 45 cm. tall.</td><td>"herba metralis".</td></tr>
<tr><td>3. Indumentum on stems and inflorescence ± spreading.</td><td>"caulibus ramisque-pedunculo rhachique- -adpresse pilosis".</td></tr>
<tr><td>4. Pedicels up to 4 mm. long.</td><td>"pedicellis 3–7 mm. longis".</td></tr>
<tr><td>5. Calyx teeth about as long as tube.</td><td>"calyx tubo 0·75 mm., dentibus 1·5 mm. longis".</td></tr>
</table>

E.6. I. lobata *Gillett* nom. nov.

Microcharis latifolia Benth., Tr. L. Soc. **25**, 297 t. 33 B (1866); L.T.A. p. 256, non *Indigofera latifolia* Micheli.

KENYA, K7 Rawlins 74; TANGANYIKA, T3 *Peter* K1135, K1137, *Drummond & Hemsley* 3146, *Mrs. Faulkner* 1262, 1671; T6 Msazani *Vaughan* 2840 BM; T7 *Stolz* 780; ZANZIBAR, *Vaughan* 2127 BM, *Hildebrandt* 938 K, BM; PEMBA, *Vaughan* 620; PORT. E. AFR., Luabo, mouth of Zambezi, *Kirk* s.n. holotype; near Quelimane *Scott* s.n., *Mrs. Faulkner* K1; Maganja da Costa *Torre* 5741.

E.7. I. microcharoides *Taub.* in Engl. Pflwelt. O. Afr. C. 209 (1895); L.T.A. No. 37.

Rhynchotropis curtisiae Johnst., Contr. Gray Herb. n.s. **75**, 23 (1925); L.T.A. p. 167.

KENYA, K4, Chyulu North, *Bally* 7673 EA; K6 Mau range S.W. of Kijabe *Curtis* 814 GH holotype of *R. curtisiae, Bally* 8257; area N. of Lake Magadi *Bogdan* 888 K, EA, 1627, 3115, *Bally* 5179; K7 Kwale distr. *Drummond & Hemsley* 3784; TANGANYIKA, T1 Musoma, *Moore* 12 EA; T2 (probably) *Fischer* 288 (B† holotype) BM sketch; Mbulu distr. *Burtt* 2101 BM; T3 Pare distr. *Haarer* 1355 K neotype EA isoneotype; Lushoto distr. *Peter* K65, *Drummond & Hemsley* 2696.

Mr. Milne-Redhead compared *Haarer* 1355 K with the holotype in 1934 and notes that they agree very well. Johnstone l.c. records the basal appendages to the anthers, which had not previously been observed in **Indigofera** or **Microcharis,** and, on this account, described *Curtis* 714 as a species of **Rhynchotropis.** However, while there is certainly an affinity between subgen. **Microcharis** and **Rhynchotropis,** the present species, in all relevant characteristics agrees with the former rather than the latter group.

131

E.8. **I. stipulosa** *Chiov.* in Cufod. Miss. Biol. Boran. 75 fig. 13 (1939).

Ethiopia, Neghelli, *Cufodontis* 228 FI holotype. 39 km. E. of Neghelli, on rocky hillside, 1350 m., 24.9.1953, *Bally* 9308.

E.9. **I. hutchinsoniana** *Gillett*, Kew Bull. **1955,** 584 (1956).

Microcharis tenella Benth., Tr. Linn. Soc. **25,** 297 t. 33 A (1866); L.T.A. p. 256 p.p. excl. syn: non *Indigofera tenella* Vahl ex DC.
"*Indigofera welwitschii Bak.*" sensu Hutch. & Dalz., F.W.T.A. **1,** 393, non Bak.
Nigeria.

E.10. **I. disjuncta** *Gillett*, Kew Bull. **1955,** 584 (1956).

α. var. **disjuncta.**

I. acutifolia Schinz, Abh. Bot. Ver. Brand. **30,** 163 (1888); non Schlecht. (1838).
"*I. arenaria A. Rich.*" sensu Bak., F.T.A. **2,** 79 (1871) et auct. omn.; non A. Rich. (1847); nec. *I. arenaria* E. Mey (1836).
I. semhaensis Vierhapper Denkschr. Ak. Wien. **71,** 362 (1907) p.p. non quoad typum.
I. welwitschii Bak. var. *simplicifolia* Bak.f., J. Bot. **70,** 253 (1932).
"*I. engleri Bak.f.*" sensu Sauvage, Rec. Bot. Miss Anti Acr. du Sahara Occ., non Bak.f.

Hejaz; Mauritania; Fr. Niger Col.; Fr. Chad. Terr.; A.-E. Sudan; S.W. Afr.; N.W. Cape Col.

β. var. **fallax** *Gillett* ibid., 585.

I. leptocarpa Hochst. & Steud. ex Balf. f., Proc. Roy. Soc. Edin. **11,** 510 (14) (1882) p.p., non E. & Z. (1836).
Socotra.

E.11. **I. phyllogramme** *Viguier*, Not. Syst. **13,** 369 (1949).

α. var. **phyllogramme.**

Madagascar, *Perrier* 474 isosyntype, *Afzelius* s.n., *Baron* s.n.

β. var. **aphylla** *Viguier* l.c.

Madagascar *Perrier* 13845 P holo., K isotype.

Perennial; leaves simple, cylindrical, up to 10 cm. long, not divided into petiole and lamina. Probably a distinct species.

E.12. **Indigofera pseudo-indigofera** (*Merxmueller*) *Gillett* comb. nov.

Microcharis galpinii N.E. Br., Kew Bull. **1897,** 258; Burtt Davy Man. Flor. Transvaal **2,** 375, non *I. galpinii* N.E. Br., Kew Bull. **1925,** 158.
Microcharis pseudo-indigofera Merxmueller, Tr. Rhod. Sci. Ass. **43,** 22 (1951).

Port. E. Afr., Lebombo mts. Swaziland border, *Exell. Mendonça & Wild* 497 BM; S. Rhodesia, Marandellas distr., *Dehn* 627 (M holotype) SRGH isotype; Salisbury distr. *Davies* 497 K, SRGH; Matobo distr. *Miller* 1628 K, SRGH, *Martineau* 203 SRGH, *West* 2544; Transvaal, Kruger park, *Codd* 6065, Komati poort *Schlechter* 11820, Lydenberg distr. *Wilms* 325, Nelspruit *Breyer* in TRV 17714, *Codd* 6474, near Barberton *Galpin* 1315 holotype of *M. galpinii*.

Merxmueller states that apical appendages are absent from the anthers of this species. In the specimens now dissected (*Davies* 497, *Codd* 6065) minute translucent apical appendages were present on the anthers of the 4 shorter stamens, which also had basal appendages about as long as themselves. Both basal and apical appendages were absent, or very much reduced, on the 5 other anthers. The species varies a good deal in the size and shape of the leaflets. The type of *M. galpinii*, and *Codd* 6065, 6474 show only unifoliolate leaves but I can detect no other difference between these and the remaining specimens, which show both 1- and 3-foliolate leaves. Further collections may show this species, which though normally perennial seems sometimes to flower in the first year, not to be specifically distinct from *I. annua* Milne-Redhead but, until further material is available it seems advisable to keep it separate.

E.13. I. : Annua *Milne-Redhead*, Kew Bull. **1936,** 470.

S. RHODESIA, Wankie, April 1932, *Levy* 36 (PRE holotype) K, SRGH isotypes.

E.14. I. contorta *Gillett* sp. nov.

Ab *I. tritoide* Bak. legumine toruloso-circinnata, seminibus aurantiacis reniformibus differt.

Herba, an perennis?, ramis prostratis subcostatis dense argenteis, pilis apice ± patentibus. *Folia* digitatim 3-foliolata ubique dense argentea, pilis ± patentibus, stipulis anguste deltoideis *c*.3 mm. longis, petiolo ad 2–3 mm. longo, foliolis elliptico-obovatis ad 5 mm. longis, 2·5 mm. latis. *Racemi c*.8–12 flori, argentei, ad 4 cm. longi, pedunculo *c*.4 mm. longo incluso, bracteis persistentibus *c*.1 mm. longis, pedicellis pilosis *c*.2 mm. longis. *Calyx* pilosus, tubo 1·5 mm. longo, dentibus deltoideo-acuminatis 1 mm. longis. *Vexillum* glabrum, late cuneato-obovatum, 5 mm. longum, 4·5 mm. latum, apice rotundatum. *Alae* glabrae, 4·3 mm. longae, unguiculo 0·7 mm. longo incluso. *Carina* glabra, rostrata, haud calcarata, 5 mm. longa, unguiculo 1·5 mm. longo incluso. *Stamen* superius liberum sterile, sine anthera; filamenta alia 3 mm. longa, antheris late ellipticis apice 0·05 mm. apiculatis, basin squamis hyalinis praeditis, eis staminum quatuor breviorum 0·3 mm. longis, eis staminum longiorum 0·1 mm. longis, antheris ipsis 0·5 mm. longis. *Ovarium* 2 mm. longum, stylo basin paulo complanato, 0·5 mm. longo, stigmate magno, discoideo-cupuliforme, 0·4 mm. diametro. *Legumen* dense argenteo-pilosum, pilis apice patentibus, contortum, subtorulosum, *c*.15 mm. longum, 1·4 mm. latum, 1 mm. crassum, *c*.7-spermum. *Semina* aurantiaca, reniformia, 1·8 mm. longa, 0·8 mm. lata, 0·5 mm. crassa.

HADHRAMAUT, Riyan airport, on coast *c*.49°15′E, 14°40′N, on sand, *K. M. Guichard* KG/HAD/349 BM holotype.

E.15. I. tritoides *Bak.*, Kew Bull. **1895,** 214 (1895); L.T.A. No. 47.

α. var. **tritoides.**

I. leptocarpa Hochst. & Steud. ex Balf. f., Proc. R. Soc. Edin. **11,** 510 (14) (1882) (1882) pro parte (quoad typum et descriptionem, non quoad plantas socotranas) non *I. leptocarpa* Eckl. & Zeyh. (1836).

I. tenuisiliqua Schwnf., Bull. Herb. Boiss **4,** app. **2,** 241 (1896), L.T.A. No. 50.

"*I. spinosa Forsk. var. spiniflora*" sensu Schweinf. Bull. Herb. boiss, **4,** app. **2,** 237 p.p. teste Schwarz, Flor. Trop. Arab. 101.

HEJAZ, *Schimper* (1837) 771 isosyntype of *I. leptocarpa* Hochst. & Steud. ex Balf. f. and *I. tenuisiliqua*, etc.; YEMEN; A.-E. SUDAN, *Schweinfurth* (1868) 297 isosyntype of *I. tenuisiliqua*, etc.; ERITREA; BR. SOMALILAND, *Miss Edith Cole* s.n. holotype, etc.

var. **obbiadensis** (*Chiov.*) *Gillett* comb. nov.

I. sesquijuga Chiov. var. *obbiadensis* Chiov., Fl. Somala **1,** 139 (1929).

SOMALIA, Obbia, in dunes, *Puccioni & Stefanini* 387 (439) FI holotype.

The status of this plant is obscure. The small leaflets and short (up to 1 cm. long) 1–3 flowered inflorescences distinguish it from var. *tritoides*. It is not impossible that these characteristics are the result of modifications due to the environment. It could be confused with *I. disjuncta* var. *fallax* but for its larger flowers, leaves all, or nearly all, 3-foliolate and the longer pedicel. The fruit is unknown. *I. tritoides* in general is a variable species, particularly in British Somaliland, and requires further investigation.

E.16. I. cufodontii *Chiov.* in Cufod. Miss. Biol. Boran. 71 fig. 12 (1939).

ETHIOPIA, South, Javello, *Cufodontis* 473, 613 FI syntypes; Mega *Gillett* 14310 K, EA; KENYA, K1, Furroli, *Gillett* 13943 K, EA; K4, Nairobi-Thika, *Verdcourt* 483, *Bogdan* 2859, 3123; Yatta plateau, 32 km. E. of Donyo Sabuk, *Bogdan* 3375; Emberre, Tana plains, *Orde Brown* s.n. BM; TANGANYIKA, T1, Musoma, *Tanner* 1694.

This species is closely related to *I. tritoides* on the one hand and to *I. microcharoides* on the other.

E.17. I. praetermissa *Bak.f.*, L.T.A. No. 92 p. 121 (1926).

"*I. welwitschii Bak.*" sensu Cronquist, F.C.B. **5**, 171 p.p. non Bak.

OUBANGUI CHARI, Crampel, *Chevalier* 6345; A.-E. SUDAN, Bongoland; *Schweinfurth* 2528; B. CONGO, Lake Moero, *Kässner* 2809 holotype; N. RHODESIA, nr. Abercorn, *Mrs. Richards* 721, 4747, 5368/A.

This differs from *I. welwitschii* in the leaves, except for the earliest simple ones, being constantly digitately 3-foliolate, and in the petioles shorter on the average. In addition the fruits of *Schweinfurth* 2528 (the only specimen with mature fruits) are substipitate.

E.18. I. welwitschii *Bak.*, F.T.A. 2, 84 (1871); L.T.A. No. 96; F.C.B. **5**, 171 pro maj. parte.

α. var. **welwitschii**.

I. variabilis De Wild., Ann. Mus. Cong. Belg. Bot. ser. **5**: **1**, 134 (1904); non N. E. Brown (1903).

B. CONGO, *Devred* 126, *Gillet* 734 BR holotype of *I. variabilis* De Wild.; A.-E SUDAN, Yei distr., *Jackson* 2077; UGANDA, West Nile, *Hazel* 695; NYASALAND, *Whyte* s.n.; N. RHODESIA, Mwinilunga distr., *Milne-Redhead* 4268; ANGOLA, *Faulkner* A288, *Welwitsch* 2031, 2032, 2066, 2067 isosyntypes.

β. var. **remotiflora** (*Taub. ex Bak.f.*) *Cronquist*, F.C.B. **5**, 172 (1954).

I. remotiflora Taub. ex Bak.f., L.T.A. No. 91 p. 121 (1926) including *I. remotiflora* var. *angolensis* Bak.f. ibid.

NIGERIA, South-West, Idanre hills, *Keay* in FHI 20247 FHI, 22593 K, FHI; North, Bauchi plateau, *Lely* 610 FHI, P 504 K, FHI; CAMEROONS, *Zenker* 486 BM Neotype (*Staudt* 486 B† holotype); B. CONGO (fide *Cronquist*, but see note); TANGANYIKA, T6, Mahenge distr., *Schlieben* 1959; T7 Iringa *Lynes* 250, Lufembe, *Schlieben* 546, Kyimbila, *Stolz* 775; T8 *Schlieben* 6392 BM; PORT E. AFR., nr. Ibo, *Stocks* s.n.; NYASALAND, *Buchanan* 206, 483, 699, *Whyte* s.n.; ANGOLA, *Gossweiler* 4820 BM type of var. *angolensis*, 9824, *Monteiro* s.n.

Stolz 775 K was named *I. remotiflora* Taub. in Berlin while Taubert was active there. Dr. Cronquist's description of this variety agrees well with the plants here listed. On the other hand, of the 3 specimens which he cites, *Kässner* 2809 is the type number of *I. praetermissa* and the Kew sheet of *Kässner* 3009, which is here described as *I. brevistaminea*, is even more distinct from var. *remotiflora*.

E.19. **I. medicaginea** *Welw. ex Bak.*, F.T.A. **2**, 86 (1871); sensu Bak. L.T.A. No. 89 et Cronquist, F.C.B. **5**, 171 pro parte, quoad typum solum.

Angola, Pungo Andongo, *Welwitsch* 2025 K, BM, 2026 BM isosyntypes.

E.20. **I. butayei** *De Wild.*, Ann. Mus. Cong. ser. **5**: **1**, 132 (1904).

"*I. medicaginea Welw. ex. Bak.*" sensu Bak.f., L.T.A. No. 89, et Cronquist, F.C.B. **5**, 171 pro parte, non Bak.

Oubangui-Chari, *Chevalier* 5531, 6517 P; B. Congo, *Butaye* 1208 BR holotype, *Sapin* s.n.; Uganda, W. Nile, *Purseglove* 1038, *Hazel* 682, *Chandler* 276; N. Rhodesia, Abercorn distr., *Mrs. Richards* 5271.

Closely resembles *I. medicaginea* which differs in having longer peduncles, much longer pedicels and shorter calyx lobes.

E.21. **I. wajirensis** *Gillett* sp. nov.

I. microcharoides, *I. cufodontii* et *I. tritoides* affinis; foliis pinnatim trifoliolatis differt.

Herba perennis, radice lignoso, caulibus patentibus, pilis medifixis, patentibus, indutis. *Folia* saepissime pinnatim trifoliolata, inferiora interdum unifoliolata, superiora rare 5-foliolata, ubique pilis medifixis $\pm$ patentibus induta, stipulis lanceolatis-attenuatis, ad 7 mm. longis et 1·5 mm. latis, petiolis 5–14 mm. longis, rhachi ultra foliola lateralia foliorum 3-foliolatorum 3–7 mm. prolongata, ultra foliola lateralia ulteriora foliorum 5-foliolatorum haud prolongata, petiolulis flavis 0·5–1 mm. longis, foliolis ellipticis, apice subacutis, terminalibus, quam lateralibus dimidie majoribus, ad 2 cm. longis 1 cm. latis. *Racemi* multiflori laxi, ad 17 [29] cm. longi, pedunculo *c.*3 cm. longo incluso, ubique pilis medifixis patentibus conspersi, bracteis lanceolatis *c.*1·3 mm. longis persistentibus, pedicellis 4–5 mm. longis, post anthesin patentibus. *Calyx* pilis medifixis patentibus indutus, campanulatus, tubo 1·7 mm. longo, dentibus deltoideo-subulatis, 1·3 mm. longis. *Vexillum* glabrum late orbiculare, 6 mm. latum, 5·5 mm. longum, unguiculo 0·5 mm. longo incluso. *Alae* glabrae, suboblongae, basin auriculatae, apice rotundatae, basin *c.*1 mm., apicem versus 1·7 mm. latae, in tota 5·7 mm. longae, unguiculo 0·8 mm. longo incluso. *Carina* glabra, apice bene rostrata, laterale paulo gibbosa sed haud calcarata, in tota 5·7 mm. longa, unguiculo 1·5 mm. longo incluso. *Stamen* vexillare liberum, aliis brevius, sterile, sine anthera, 1·8 mm. longum; stamina alia filamentis 2·5–3 mm. longis, antheris ellipticis, apice apiculatis, basin squamo hyalino deltoideo appendiculatis, squamo staminum 4 breviorum illis 5 longiorum maiore, anthera ipsa 0·5 mm., apiculo 0·05 mm., squamo basale 0·15–0·2 mm., longo. *Ovarium* dense strigosum, 2 mm. longum, multiovulatum, stylo glabro 1·2 mm. longo ab ovario ad angulum 140° producto, basin valde compresse dilatato, stigmate cupuliforme, inclinato, 0·5 mm. diametro. *Legumen* paulo immatuum rectum, pilosum, pilis medifixis plus minusve patentibus, ad rhachidem inflorescentiae plus minusve parallelum, cum pedicello angulum 90° praebens, ad 27-spermum, 35–40 [–50] mm. longum, 1·3 mm. latum, 0·6 mm. crassum. *Semina* matura haud visa.

Kenya, K1 Wajir, 240 m., rainfall 200 mm. with maxima in April and November, red sandy loam, *Commiphora-Acacia* open scrub with much *Delonix elata*, overgrazed on account of wells, corolla brick red, 27.5.1952, *Gillett* 13368 K holo. EA isotype ibid. 21.6.1951, *Kirrika* 58 K, EA.

E.22. **I. asparagoides** *Taub.* in Engl. Pflwelt O.Afr. C. 210 (1895); L.T.A. No. 106; F.C.B. **5**, 172.

Gaboon (fide Cronquist); B. Congo, East and South-East; Uganda, U2–4; Kenya, K5; Tanganyika, T1, Bukoba (*Stuhlmann* 3989 B†, 4101 B† syntypes), Ihangiro (*Stuhlmann* 3353 B† syntype) T4. T7; N. Rhodesia, Abercorn distr., Mwinilunga distr.

E.23. **I. tisserantii** (*Pellegr.*) *Pellegr.*, B. Soc. Bot. Fr. **94**, 6 (1947); F.C.B. **5**, 173.

I. asparagoides Taub. var. *tisserantii* Pellegr., B. Soc. Bot. Fr. **72**, 538 (1925).
OUBANGUI, Waka region, Bayedou rocks, *Tisserant* 1650 P holotype.

RHYNCHOTROPIS *Harms*, Engl. B.J. **30**, 86 (1901).

Type species *R. poggei* (Taub.) Harms.

Centre of distribution S. Tropical Africa.

Calyx *c.*4·5–5 mm. long; carina 8–10 mm. long, with a long beak; filaments 5–6 mm. long; inflorescences either terminal or in leaf axils, always on leafy shoots; leaves often over 3 mm. wide; ovules 5–6 . . . . 1. *R. poggei.*

Calyx *c.*3–3·5 mm. long; carina 6–7 mm. long, with a short beak; filaments 3·5–4·5 mm. long; inflorescences often on short precocious shoots without normal leaves; leaves not over 3 mm. wide; ovules 3–5 . . . . 2. *R. marginata.*

1. **R. poggei** (*Taub.*) *Harms*, Engl. B. J. **30**, 86 (1901); L.T.A. p. 166; F.C.B. **5**, 174 pro maj. parte.

Indigofera poggei Taub. in Engl. & Prantl. Pfanzenfam. **3**: **3**, 260 (1894) fig. 115 H-K; Engl. B.J. **26**, 284 (1899).

R. dekindtii Harms, Engl. B.J. **30**, 87 (1901); L.T.A. p. 167.

B. CONGO, Katanga, Kundelungu Mts., *Kässner* 2617 BM, 2715 K, BM, Tshinsenda *Rogers* 10107 K, BM, 10394 K, BM. N. RHODESIA, North Prov. Luwingu, *Japp* 47 BM, 64 km. S. of Abercorn, *Miss Ricardo* 111 BM, N. of Abercorn *Mrs. Richards* 5576; South. Prov. Mumbwa *Miss Macaulay* 1133; West Prov. Solwezi-Kasempa, *Greenway & Brenan* 8134, Mwinilunga *Milne-Redhead* 905. ANGOLA (*Pogge* B† holotype), (*Dekindt* 93 B†, 94 B† syntypes of *R. dekindtii*), *Baum* 742, K, BM, *Gossweiler* 2003 BM, 2796 BM, 3050 K, BM, *Young* 712 BM, 1015 BM, 1398 BM.

R. dekindtii was distinguished on the basis of its broader leaves, but there seems to be an uninterrupted series from plants with leaves under 2 mm. wide to plants such as *Milne-Redhead* 905 with leaves up to 10 mm. wide.

2. **R. marginata** (*N. E. Brown*) *Gillett* comb. nov.

Crotalaria marginata N. E. Brown, Kew Bull. **1921**, 291.

R. praecox Bak.f., L.T.A. 167 (1926); F.C.B. **5**, 174.

B. CONGO, Katanga, Elizabethville, *Rogers* 26286 syntype 26288 holotype, Baya station *Burtt Davy* 18079 BM type of *R. praecox*; N. RHODESIA, Luanshya, *Fanshawe* 1399, 2443, Chingola *Fanshawe* 2458, 2511.

Rogers 26288 clearly shows remains of precocious flowering shoots arising from the base of the stem, while it also bears other inflorescences among the leaves. The small flowers show its affinity with *R. praecox* and there can be little doubt that it is but an example of this species in which the leafy shoots have grown out and produced further inflorescences.

ECONOMICS

1. **Dyes** *I. articulata* Gouan, *I. coerulea* Roxb, *I. tinctoria* L., *I. arrecta* Hochst. ex A. Rich. *I. suffruticosa*, Mill. *I. guatimalensis* Moc. & Sessé ex Prain & Bak.f.—and perhaps a few other species to a lesser extent, were once of importance as the source of *Indigo*. Their cultivation still continues here and there on a small scale

but is no longer of importance in world trade. The best readily available summary of the history of the natural *Indigo* industry seems to be that given by Burkill in " Economic products of the Malay peninsula " (1935).

2. **Medicine:** Minor uses of various species of *Indigofera* in native medicine are reported in the standard works for China (Stuart), Malaya (Burkill), India (Watt), West Africa (Dalziel) and South Africa (Watt and Brandwijk).

3. **Poison:** (a) The following species have been shown to be definitely toxic even in relatively small amounts.

I. australis Willd., Australia, Hurst Poisonous Pl. N.S. Wales (1942).

I. boviperda Morrison, Australia. J. Bot. **50,** 166 (1912).

I. galegoides DC, S.E. Asia, stated by Pammel, Man. Pois. Pl. to contain HCN.

(b) The following species have been recorded as perhaps, or sometimes, poisonous without this having been proved by any thorough investigation.

I. hirsuta L "suspected in Queensland" Hurst l.c.

I. linifolia (L.f.) Retz (as "I. linophylla" in error) ibid.

I. tinctoria L. Pammel l.c.

I. suffruticosa Mill (*I. anil*) ibid.

I. cryptantha Benth. ex Harv. Watt & Brandwijk, medicinal and poisonous plants of South Africa (1932).

I. hololeuca Benth. ex Harv. ibid.

I. hilaris Eckl. & Zeyhr. ibid.

of these the first two have also been reported as useful fodder plants.

(c) The following species, originally thought innocuous, have been shown on investigation to be poisonous in certain forms, under certain growth conditions or when eaten in really large amounts, as may happen when other food is scarce.

I. enneaphylla L. responsible, when eaten in large quantities, for the "Birdsville" disease of horses which causes considerable financial losses in West Australia, see Austr. Vet. J. 27, 185–8 (1951). It is interesting that this species which is also common in India is not there recognised as poisonous. Indeed no species of *Indigofera* is listed in Chopra Badhwa and Gosh. "The poisonous plants of India" (1949). The species has also been recorded as a useful fodder plant both in Queensland Agr. J. 60, 5–13 (1945) and in India (Watt).

I. sessilifolia DC (*I. patens* Eckl. & Zehr.) Van der Walt & Steyn, Onderstepoort J. Vet. Sci. **12,** 335–66 (1939). Test sheep died but from Tympanites (bloating) a condition sometimes caused merely by an excess of good fodder.

I. spicata Forsk. (*I. hendecaphylla* Jacq.). This species presents a particularly difficult problem as it has been repeatedly investigated in several parts of the tropics and found to be one of the most valuable leguminous fodder plants for several types of tropical pasture. As such its distribution is being actively extended. However as far back as 1926 A. E. Haarer recorded on his specimen B 135 in the Kew Herbarium, from Arusha Tanganyika "said to cause abortion in cattle". This note was published by Dalziel in "Useful Plants of W. Trop. Afr." (1937). Some of the numerous investigations of the plant which have been published since then are summed up by Payne & Naidu, Agr. J. Fiji 26, 1–3 as follows:

Emmel & Ritchey, J. Am. Agron, **33,** 675 (1941)—toxic to rabbits (in Florida).

Nordfelt & Young, Hawaii Agr. Exp. sta. Progr. notes 55 (1949)—reproductive disturbances in cattle.

Rosenberg & Palafox, World Poultry Sci. J. **6,** 284 (1950)—toxic to chickens in Hawaii.

Nordfelt et al, Hawaii Agr. Exp. Sta. Bull. 15 (1952)—if eaten in large amounts causes abortion in cattle and death in sheep and rabbits.

Jaganathan, Trop. Agr. 109, 297 (1953) Ceylon—a slight retardation in growth when fed in large amounts to calves.

Warmke & Freyre, Ann. Rept. Fed. Exp. Sta. Puerto Rico (1953)—death in Guinea pigs.

Payne and Naidu could find no clear ill effects under certain conditions in Fiji but recommend a suspension of all further propagation of the species until it has been shown that strains can be selected not liable to be poisonous under any conditions likely to occur.

4. Fodder: The following have been reported as valuable fodder species giving satisfactory yields under experimental cultivation.

(a) Perennials:

I. spicata Forsk. (*I. endecaphylla* Jacq.) in Florida, Georgia, Puerto Rico, Ceylon, Kenya, Fiji, etc.

I. subulata Vahl ex Poir, Puerto Rico, etc.

I. schimperi Jaub. & Spach. (*I. tettensis*), Kenya.

I. oblongifolia Forsk. (*I. paucifolia* Del.), India (Watt).

I. pilosa Poir., Florida.

(b) Annuals:

I. hirsuta L. Florida, Georgia, Brazil.

5. Cover crops and green manures, etc.: The following have been recommended.

I. spicata Forsk. (*I. endecaphylla* Jacq.) in Malaya, Ceylon, Guatemala and elsewhere. As a creeping plant this species has been recommended for the control of erosion in Ceylon by planting on banks and "bunds".

I. hirsuta L., *I. arrecta* Hochst. ex A. Rich, *I. suffruticosa* Mill., *I. tinctoria* L. (*I. sumatrana*).

6. Human food: Watt records that the seeds of the following Indian species are used in times of famine. *I. cordifolia* Heyne ex Roth, *I. glandulosa* Willd, *I. linifolia* (L.f.) Retz.

GEOGRAPHICAL DISTRIBUTION

The table following this note summarises the area of distribution of the taxa treated in this paper. The list is intended to be complete for those countries marked *. Figures giving the number of species of a group in a particular area are enclosed in brackets where there has not been an attempt to make them as complete as possible. The following points of interest emerge.

A. *The scarcity of the genus in areas of equatorial forest*

No species are recorded from Fernando Po and but two from Spanish Guinea. Even such species as are found in regions dominated by equatorial forest are usually weeds or relics of cultivation.

B. *The extraordinary absence of the genus from the Mediterranean region*

In this and the adjacent Irano-Turanian region not one species is native if *I. oblongifolia* Forsk from the "Sudano-deccanian" enclave by the Dead Sea be excepted. It would seem that this absence cannot be explained by any inability of the genus to adapt itself to Mediterranean conditions. The genus is for example strongly represented in those parts of the Cape Province which have a climate of

Mediterranean type, there being no less than 17 species on the Cape peninsula, some of which extend up the mountains to at least 600 m. In E. Africa the genus reaches at least 3500 m. In the Himalayas and China the genus is well represented; at least two species from this region, *I. kirilowii* and *I. amblyantha* being hardy in cultivation in New England. Several species are native in the U.S.A. In Australia *I. australis* reaches to Tasmania. It seems that here there is a mystery for "historical plant geography" to solve.

C. *Disjunct areas of distribution N. and S. of the equator*

The table below lists instances where a widely disjunct area, one portion in the subarid north, the other in the subarid south, is occupied by a species, varieties of a species or by pairs of vicarious species.

North	*South*
I. trigonelloides J. & S.	*I. trigonelloides*
I. disjuncta Gillett	*I. disjuncta*
I. pilosa Poir.	*I. pilosa var. angolensis*
I. diphylla Vent.	*I. flavicans* Bak.
I. senegalensis Lam.	*I. praticola* Bak.f.
I. aspera Perr. ex DC.	*I. charlieriana* Schinz
I. spinosa Forsk.	*I. longispina* Gillett
I. berhautiana Gillett	*I. arenophila* Schinz and *I. eylesiana* Gillett
Cyamopsis senegalensis G. & P.	*C. serrata* Schinz.

This type of discontinuous distribution is not uncommon. Cases occur in *Kirkia, Moringa, Krauseola, Zygophyllum*. It is parallel to the discontinuous distribution of genera between the Mediterranean and Cape regions, e.g. *Secale, Trifolium* section *Lagopus*. It may be explained on the supposition that there have always been subarid areas in North and Southern Africa throughout the life of the genus while the climate in E. Africa has been unstable. Species which crossed the equator during a previous arid phase in E. Africa would have been wiped out there on the return of more humid conditions, and would not necessarily reinvade East African arid areas when these once more appeared. An alternative (or supplementary) hypothesis is that the equatorial belt may have been crossed by the temporary colonisation of unstable local open habitats. Thus *I. charlieriana* Schinz var. *sessilis* (Chiov.) Gillett seems to have crossed the equator from south to north on both sides of Africa on coastal sands. *Lotus arabicus* L. in the north is probably not separable from *L. mossamedensis* Welw. ex Bak. which occupies a widely separate area in the south. Recently (*Bullock* 3461) *L. arabicus* has been collected for the first time in East Africa on the exposed shores of a receding lake.

CYAMOPSIS

INDIGOFERA

A. **Acanthonotus**

B. **Amecarpus**

	Mauritania*	Senegal and Gambia*	Fr. and Port. Guinea*	Sierra Leone and Liberia*	Ivory Coast*	Fr. Sudan and Upper Volta*	Gold Coast and Br. Togo*	Fr. Togo and Dahomey*	Fr. Niger Colony*	Nigeria and Br. Cameroons*	Fr. Equatorial Afr.	Belgian Congo	A.-E. Sudan and Egypt	Arabia	Asia east of Arabia	Australia	Eritrea	Ethiopia
CYAMOPSIS																		
1 C. tetragonoloba															?			
2 C. senegalensis		×				×							×	×			×	
3 C. serrata																		
INDIGOFERA																		
A 1 nummulariifolia		×	×			×	×			×	×	×	×		×			
A 2 drepanocarpa										×	×	×						
A 3 cuitoensis																		
B 1 letestui											×		×					
B 2 circinnata																		
B 3 sessilifolia																		
B 4 rautanenii																		
B 5α rhytidocarpa ssp. r.																		
B 5β ,, ssp. angolensis																		
B 6 leendertziae																		
B 7 holubii																		
B 8α torulosa var. t.																		
B 8β ,, var. angustiloba																		
B 9 oligophylla																		
B 10 arabica														×			×	×
B 11 chirensis																	×	×
B 12 hochstetteri						×			×	×		×	×	×	×		×	×
B 13 senegalensis		×				×				×	×							
B 14 praticola																		
B 15 sisalis																		
B 16 wilmaniae																		
B 17 dauensis																		?
B 18 lasiantha																		
B 19 fanshawei																		
B 20α charlieriana var. c.																		
B 20β ,, var. scaberrima																		
B 20γ ,, var. sessilis								×										
B 20δ ,, var. lata																		
B 21 aspera		×				×	×			×			×					
B 22 demissa												×						
B 23 concinna																		
B 24 bussei																		
Total Amecarpus	−	2	−	−	−	3	1	1	1	3	2	2	3	2	1	−	3	4

Taxon	Code	Somalia	Uganda*	Kenya*	Tanganyika*	Zanzibar and Pemba*	Madagascar	Port. E. Africa*	Nyasaland*	S. Rhod. and Trop. Bechuanald*	N. Rhodesia*	Angola	S.W. Afr. and N.W. Cape	N. Central Cape and Extra Trop. Bechuanald	Transvaal and Swaziland	Orange Free State and Basutoland	Natal	Eastern Cape	America
Cyamopsis																			
C. tetragonoloba	1																		
C. senegalensis	2																		
C. serrata s. lat.	3									×		×	×	×					
Indigofera																			
nummulariifolia	A 1				×		×	×		×	×								
drepanocarpa	A 2		×	×	×	×					×								
cuitoensis	A 3											×							
letestui	B 1																		
circinnata	B 2													×	×				
sessilifolia	B 3												×	×			×	×	
rautanenii	B 4												×						
rhytidocarpa ssp. r.	B 5α									×					×	×	×		
,, ssp. angolensis	B 5β											×	×						
leendertziae	B 6														×				
holubii	B 7							×					×	×	×				
torulosa var. t.	B 8α																×	×	
,, var. angustiloba	B 8β														×				
oligophylla	B 9									×									
arabica	B 10																		
chirensis	B 11																		
hochstetteri	B 12		×	×	×														
senegalensis	B 13			×	×					×									
praticola	B 14									×				×					
sisalis	B 15												×	×					
wilmaniae	B 16			×															
dauensis	B 17											×							
lasiantha	B 18											×							
fanshawei	B 19									×	×	×	×	×					
charlieriana var. c.	B 20α							×		×	×		×	×					
,, var. scaberrima	B 20β												×	×					
,, var. sessilis	B 20γ	×		×	×	×		×				?			×				
,, var. lata	B 20δ									?			×						
aspera	B 21			×	×			×	×	×	×								
demissa	B 22				×			×		×	×								
concinna	B 23				×														
bussei	B 24				×														
Total Amecarpus		1	1	4	7	2	–	5	1	7	4	4	8	7	6	2	2	2	–

C.1. **Latestipulatae**

C.2. **Paniculatae**

C.3a. *Juncifoliae*

	Mauritania*	Senegal and Gambia*	Fr. and Port. Guinea*	Sierra Leone and Liberia*	Ivory Coast*	Fr. Sudan and Upper Volta*	Gold Coast and Br. Togo*	Fr. Togo and Dahomey*	Fr. Niger Colony*	Nigeria and Br. Cameroons*	Fr. Equatorial Afr.	Belgian Congo	A.-E. Sudan and Egypt	Arabia	Asia east of Arabia	Australia	Eritrea	Ethiopia
C.1. 1 ischnoclada																		
,, 2 gairdnerae																		
,, 3 arenophila																		
,, 4 kuntzei																		
,, 5 faulknerae																		
,, 6 eylesiana																		
,, 7 burttii																		
,, 8 inhambanensis																		
,, 9 taborensis																		
,, 10 berhautiana		×					×		×	×								
,, 11α strobilifera ssp. s.						×			×	×	×		×					
,, 11β ,, ssp. lanuginosa																		
Total Latestipulatae	−	1	−	−	−	1	1	−	2	2	1	−	1	−	−	−	−	−
C.2.a.1 polysphaera					×		×			×	×	×	×					
,, 2 trialata					×		×			×	×	×						
,, 3α paniculata ssp. p.		×	×	×	×		×			×								
,, 3β ,, ssp. gazensis																		
,, 4 macrocalyx		×	×	×	×	×												
,, 5 terminalis		×			×	×												
,, 6 megacephala			×															
,, 7 capitata		×	×	×		×				×	×		×					
,, 8 paracapitata										×	×	×						
,, 9 congesta			×	×	×			×		×	×	×	×					
,, 10 lotononoides											×		×					
,, 11 dasycephala										×	×							
,, 12 brassii						×												
,, 13 pulchra		×	×	×	×	×	×	×	×	×	×	×	×					×
,, 14 vanderystii												×						
,, 15 nigricans						×	×	×		×								
,, 16 oubanguiensis							×	×					×					
,, 17 leptoclada		×				×	×											
,, 18 bracteolata		×				×	×	×	×	×			×					
C.2.b.1α trichopoda var. t.		×																
,, 1β ,, var. oubanguiensis											×							
,, 2 nigritana		×				×	×			×	×	×	×					
,, 3 latisepala										×								
Total Paniculatae	−	9	6	5	7	9	9	5	2	12	10	7	8	−	−	−	−	1
C.3.a.1 podophylla																		

Somalia	Uganda*	Kenya*	Tanganyika*	Zanzibar and Pemba*	Madagascar	Port. E. Africa*	Nyasaland*	S. Rhod. and Trop. Bechuanald*	N. Rhodesia*	Angola	S.W. Afr. and N.W. Cape	N. Central Cape and Extra Trop. Bechuanald	Transvaal and Swaziland	Orange Free State and Basutoland	Natal	Eastern Cape	America	
			×				×											ischnoclada .. C.1. 1
									×									gairdnerae.. ,, 2
								×	×		×							arenophila.. ,, 3
			×			×												kuntzei .. ,, 4
						×	×											faulknerae.. ,, 5
								×	×									eylesiana .. ,, 6
			×															burttii .. ,, 7
						×		×					×		×			inhambanensis .. ,, 8
			×															taborensis .. ,, 9
																		berhautiana .. ,, 10
		×	×				×	×	×									strobilifera ssp. s... ,, 11α
		×	×	×		×												,, ssp. lanuginosa ,, 11β
–	–	2	6	1	–	4	3	4	4	–	1	–	1	–	1	–	–	Total **Latestipulatae**
	×		×						×	×								polysphaera .. C.2.a.1
																		trialata .. ,, 2
	×	×	×	×		×				×								paniculata ssp. p... ,, 3α
			×			×	×	×										,, ssp. gazensis .. ,, 3β
																		macrocalyx .. ,, 4
																		terminalis .. ,, 5
																		megacephala .. ,, 6
	×								×	×								capitata .. ,, 7
	×									×								paracapitata .. ,, 8
	×	×	×			×			×	×								congesta .. ,, 9
																		lotononoides .. ,, 10
																		dasycephala .. ,, 11
																		brassii .. ,, 12
			×							×								pulchra .. ,, 13
																		vanderystii .. ,, 14
																		nigricans .. ,, 15
																		oubanguiensis .. ,, 16
																		leptoclada .. ,, 17
																		bracteolata .. ,, 18
																		trichopoda var. t... C.2.b.1α
																		,, var. oubanguiensis ,, 1β
																		nigritana .. ,, 2
																		latisepala .. ,, 3
–	5	2	5	1	–	3	1	1	3	6	–	–	–	–	–	–	–	Total **Paniculatae**
						×												podophylla .. C.3.a.1

C.3.b. *Brevi-erectae*
C.3.f. *Brevipatentes*
C.3.g. *Pilosae*

Columns 1–10 (Mauritania – Nigeria and Br. Cameroons)

	Mauritania*	Senegal and Gambia*	Fr. and Port. Guinea*	Sierra Leone and Liberia*	Ivory Coast*	Fr. Sudan and Upper Volta*	Gold Coast and Br. Togo*	Fr. Togo and Dahomey*	Fr. Niger Colony*	Nigeria and Br. Cameroons*
C.3.b.1 linifolia										
„ 2 cordifolia	×								×	
„ 3 enneaphylla										
„ 4 trigonelloides										
„ 5 sessiliflora	×	×			×				×	
„ 6 tetrasperma						×	×	×		
„ 7 simplicifolia		×	×	×						×
„ 8 leprieurii		×	×	×			×	×		×
„ 9 prieureana		×							×	×
„ 10α stenophylla var. s.		×	×			×	×			×
„ 10β „ var. ampla						×	×			×
„ 11 griseoides										
„ 12 mildbraediana										×
„ 13 andrewsiana										
„ 14 microcalyx										
„ 15 pobeguinii			×							
Total Brevi-erectae	2	5	4	2	1	3	4	2	3	6
C.3.f.1 ruspolii										
„ 2 ogadensis										
„ 3 suaveolens									×	
„ 4 nebrowniana										
„ 5 bainesii										
„ 6 pseudointricata										
„ 7 tanaensis										
Total Brevipatentes	–	–	–	–	–	–	–	–	1	–
C.3.g.1 petiolata										
„ 2 kirkii										
„ 3 erythrogramma										
„ 4 nummularia										
„ 5 biglandulosa										
„ 6α pilosa var. p.		×	×				×			×
„ 6β „ var. angolensis										
„ 7 fulvopilosa				×		×	×			×
„ 8 torrei										
Total Pilosae	–	1	1	1	–	1	2	–	–	2

Columns 11–19 (Fr. Equatorial Afr. – Somaliland Fr. and Brit.)

	Fr. Equatorial Afr.	Belgian Congo	A.-E. Sudan and Egypt	Arabia	Asia east of Arabia	Australia	Eritrea	Ethiopia	Somaliland Fr. and Brit.
C.3.b.1 linifolia			×				×	×	
„ 2 cordifolia			×				×	×	
„ 3 enneaphylla							×	×	
„ 4 trigonelloides								×	
„ 5 sessiliflora			×	×					
„ 6 tetrasperma									
„ 7 simplicifolia	×	×	×						
„ 8 leprieurii									
„ 9 prieureana	×	×	×						
„ 10α stenophylla var. s.	×	×	×						
„ 10β „ var. ampla									
„ 11 griseoides									
„ 12 mildbraediana	×		×						
„ 13 andrewsiana			×						
„ 14 microcalyx	×	×							
„ 15 pobeguinii									
Total Brevi-erectae	5	4	8	1	–	–	3	4	–
C.3.f.1 ruspolii								×	×
„ 2 ogadensis								×	
„ 3 suaveolens		×					×	×	
„ 4 nebrowniana									
„ 5 bainesii									
„ 6 pseudointricata									
„ 7 tanaensis									
Total Brevipatentes	–	1	–	–	–	–	1	3	1
C.3.g.1 petiolata		×							
„ 2 kirkii									
„ 3 erythrogramma		×							
„ 4 nummularia									
„ 5 biglandulosa			×						
„ 6α pilosa var. p.		×	×				×		
„ 6β „ var. angolensis									
„ 7 fulvopilosa			×						
„ 8 torrei									
Total Pilosae	–	3	3	–	–	–	1	–	–

144

Somalia	Uganda*	Kenya*	Tanganyika*	Zanzibar and Pemba*	Madagascar	Port. E. Africa*	Nyasaland*	S. Rhod. and Trop. Bechuanaland*	N. Rhodesia*	Angola	S.W. Afr. and N.W. Cape	N. Central Cape and Extra Trop. Bechuanaland	Transvaal and Swaziland	Orange Free State and Basutoland	Natal	Eastern Cape	America	
																		linifolia C.3.b.1
																		cordifolia .. .. ,, 2
																		enneaphylla .. ,, 3
											×							trigonelloides .. ,, 4
																		sessiliflora .. ,, 5
																		tetrasperma .. ,, 6
	×			×		×				×								simplicifolia .. ,, 7
																		leprieurii .. .. ,, 8
																		prieureana.. .. ,, 9
	×		×															stenophylla var. s. ,, 10α
																		,, var. ampla .. ,, 10β
			×						×	×								griseoides .. .. ,, 11
			×							×								mildbraediana .. ,, 12
	×																	andrewsiana .. ,, 13
	×		×			×	×		×	×								microcalyx .. ,, 14
																		pobeguinii.. .. ,, 15
–	4	–	4	1	–	2	1	–	2	4	(1)	–	–	–	–	–	–	Total *Brevi-erectae*
×																		ruspolii C.3.f.1
																		ogadensis .. .. ,, 2
																		suaveolens.. .. ,, 3
														×					nebrowniana .. ,, 4
								×					×					bainesii .. .. ,, 5
		×																pseudointricata .. ,, 6
																		tanaensis .. .. ,, 7
1	–	1	–	–	–	–	–	1	–	–	–	–	2	–	–	–	–	Total *Brevipatentes*
			×		×	×												petiolata C.3.g.1
			×			×	×	×	×	×								kirkii .. .. ,, 2
										×								erythrogramma .. ,, 3
			×															nummularia .. ,, 4
																		biglandulosa .. ,, 5
																		pilosa var. p. .. ,, 6α
										×								,, var. angolensis ,, 6β
	×		×			×	×	×	×									fulvopilosa .. ,, 7
						×												torrei .. .. ,, 8
–	1	–	4	–	1	4	2	2	2	3	–	–	–	–	–	–	–	Total *Pilosae*

C.3.c. *Anomalae*

C.3.d. *Dissitiflorae*

C.3.e. *Spinosae*

	Mauritania*	Senegal and Gambia*	Fr. and Port. Guinea*	Sierra Leone and Liberia	Ivory Coast*	Fr. Sudan and Upper Volta*	Gold Coast and Br. Togo●	Fr. Togo and Dahomey*	Fr. Niger Colony*	Nigeria and Br. Cameroons*	Fr. Equatorial Afr.	Belgian Congo	A.-E. Sudan and Egypt	Arabia	Asia east of Arabia	Australia	Eritrea	Ethiopia	Somaliland Fr. and Brit.
C.3.c.1 scarciesii			×																
C.3.d.1 dendroides ..		×	×	×	×	×	×	×		×	×	×	×						
,, 2 sp. aff. amitinae ..																			
,, 3 filipes																			
,, 4α heudelotii var. h...			×	×					×										
,, 4β ,, var. fairchildii			×																
,, 5 elliotii		×	×	×															
,, 6 brevicalyx.. ..												×					×	×	
,, 7 monantha												×							
,, 8 cuneata																			
,, 9 manyoniensis ..																			
,, 10 micropetala ..																			
,, 11α tanganyikensis var. t.												×						×	
,, 11β ,, var. paucijuga																			
,, 12 mwanzae																			
,, 13 elwakensis ..																			
,, 14 ambelacensis ..												×	×				×		
,, 15 kongwaensis ..																			
,, 16α wituensis var. w...																			
,, 16β ,, var. occidentalis										×									
,, 17 mendonçae ..																			
,, 18 sparsa																		×	
,, 19α tenuis var. tenuis..																			
,, 19β ,, var. major ..																			
,, 20 dissitiflora.. ..																			
,, 21 vohemarensis ..												×						×	
,, 22α congolensis var. c.		×				×				×	×	×							
,, 22β ,, var. bongensis													×						
,, 23 geminata						×		×		×									
,, 24 zavattarii																		×	
,, 25 nairobiensis ..																			
,, 26 hedyantha.. ..																			
Total *Dissitiflorae*	–	3	4	3	1	3	1	2	1	5	2	7	3	–	–	–	2	5	–
C.3.e.1 rothii																		×	
,, 2 spinosa													×	×			×	×	×
,, 3 basiflora																			
,, 4 longispina.. ..																			

Species	Socotra	Somalia	Uganda*	Kenya*	Tanganyika*	Zanzibar and Pemba*	Madagascar	Port. E. Africa*	Nyasaland*	S. Rhod. and Trop. Bechuanald*	N. Rhodesia*	Angola	S.W. Afr. and N.W. Cape	N. Central Cape and Extra Trop. Bechuanald	Transvaal and Swaziland	Orange Free State and Basutoland	Natal	Eastern Cape	America
scarciesii C.3.c.1																			
dendroides .. C.3.d.1			×	×	×	×		×	×	×	×	×							
sp. aff. amitinae .. ,, 2															×	×			
filipes .. .. ,, 3														×	×	×			
heudelotii var. heud. ,, 4α									×	×	×								
,, var. fairchildii ,, 4β																			
elliotii .. .. ,, 5																			
brevicalyx .. .. ,, 6			×	×	×														
monantha .. .. ,, 7				×	×														
cuneata .. .. ,, 8				×	×														
manyoniensis .. ,, 9					×					×									
micropetala .. ,, 10				×	×														
tanganyikensis var. t. ,, 11α			×	×	×			×											
,, var. paucijuga ,, 11β				×															
mwanzae .. .. ,, 12					×						×								
elwakensis .. .. ,, 13					×														
ambelacensis .. ,, 14			×	×	×														
kongwaensis .. ,, 15				×	×														
wituensis var. wit. .. ,, 16α					×														
,, var. occidentalis ,, 16β																			
mendonçae .. ,, 17																			
sparsa .. .. ,, 18																			
tenuis var. tenuis .. ,, 19α			×	×	×			×											
,, var. major .. ,, 19β																			
dissitiflora .. ,, 20				×	×			×		×									
vohemarensis .. ,, 21			×		×	×		×											
congolensis var. cong. ,, 22α																			
,, var. bongensis ,, 22β																			
geminata .. .. ,, 23																			
zavattarii .. .. ,, 24																			
nairobiensis .. ,, 25			×	×															
hedyantha .. .. ,, 26				×	×			×	×	×	×				×		×	×	
Total *Dissitiflorae*	–	–	7	13	16	2	–	6	3	5	4	1	–	–	–	–	–	–	–
rothii C.3.e.1				×	×														
spinosa .. .. ,, 2					×														
basiflora .. .. ,, 3																			
longispina .. .. ,, 4													×						

C.3.h. *Viscosae*

C.3.i. *Centrae*

	Mauritania*	Senegal and Gambia*	Fr. and Port. Guinea*	Sierra Leone and Liberia*	Ivory Coast*	Fr. Sudan and Upper Volta*	Gold Coast and Br. Togo*	Fr. Togo and Dahomey*	Fr. Niger Colony*	Nigeria and Br. Cameroons*	Fr. Equatorial Afr.	Belgian Congo	A.-E. Sudan and Egypt	Arabia	Asia east of Arabia	Australia	Eritrea	Ethiopia	Somaliland Fr. and Brit.
C.3.h.1 hilaris												×							
„ 2 sordida																			
„ 3 milne-redheadii																			
„ 4 heterotricha																			
„ 5 lydenbergensis																			
„ 6 curvata																			
„ 7α mimosoides var. m.										×	×	×						×	
„ 7β „ var. viscidior																			
„ 7γ „ var. brachycarpa																			
„ 8 adenoides																			
„ 9 viscidissima																			
„ 10α colutea var. c.		×				×				×	×	×	×	×			×	×	
„ 10β „ var. linearis																		×	
„ 10γ „ var. somalensis																		×	×
„ 10δ „ var. dembianensis																		×	
„ 10ε „ var. grandiflora																		×	
„ 11 barteri							×			×									
„ 12 dyeri																			
„ 13 delagoaensis																			
„ 14 hewittii																			
„ 15α bangweolensis var. b.																			
„ 15β „ var. lacus-victoriae																			
„ 16α secundiflora var. s.		×				×	×		×	×			×						
„ 16β „ var. rubripilosa											×	×	×				×	×	
„ 17 argentea	×					×					×		×	×			×	×	×
„ 18 conferta										×									
Total *Viscosae*	1	2	–	–	–	3	2	–	1	5	4	4	4	2	–	–	3	8	2
C.3.i. 1 quarrei												×							
„ 2 ugandensis																			
„ 3α subulifera var. s.												?							
„ 3β var. polysperma																			
„ 4 subargentea												×							
„ 5 microcephala												×							
„ 6 wildiana																			
„ 7 phyllanthoides																			
„ 8α vicioides var. v.												×					×	×	
„ 8β „ var. rogersii												×							
„ 8γ „ var. occidentalis											×								
„ 9 livingstoniana																			
„ 10 heterocarpa																			
„ 11 hermannioides																			
„ 12 taylori																			
Total *Centrae*	–	–	–	–	–	–	–	–	–	–	1	6	–	–	–	–	1	1	–

Somalia	Uganda*	Kenya*	Tanganyika*	Zanzibar and Pemba*	Madagascar	Port. E. Africa*	Nyasaland*	S. Rhod. and Trop. Bechuanald*	N. Rhodesia*	Angola	S.W. Afr. and N.W. Cape	N. Central Cape and Extra Trop. Bechuanald	Transvaal and Swaziland	Orange Free State and Basutoland	Natal	Eastern Cape	America	Species	
			×			×	×	×	×				×	×	×	×		hilaris	C.3.h.1
								×					×					sordida	„ 2
			×				×		×									milne-redheadii	„ 3
											×	×	×					heterotricha	„ 4
						×							×					lydenbergensis	„ 5
			×															curvata	„ 6
	×	×	×			×	×	×	×	×								mimosoides var. m.	„ 7α
			×															„ var. viscidior	„ 7β
			×															„ var. brachycarpa	„ 7γ
							×	×					×					adenoides	„ 8
		×	×			×	×	×	×	×								viscidissima	„ 9
×	×	×	×	×		×	×	×	×	×	×							colutea var. c.	„ 10α
		×	×				×	×	×									„ var. linearis	„ 10β
		×	×															„ var. somalensis	„ 10γ
	×	×																„ var. dembianensis	„ 10δ
			×															„ var. grandiflora	„ 10ε
						×	×	×	×				×					barteri	„ 11
						×		×	×				×					dyeri	„ 12
								×	×									delagoaensis	„ 13
									×									hewittii	„ 14
			×															bangweolensis var. v.	„ 15α
	×																	„ var. lacus-victoriae	„ 15β
		×	×															secundiflora var. s.	„ 16α
																		„ var. rubripilosa	„ 16β
		×																argentea	„ 17
								×										conferta	„ 18
(1)	3	6	12	1	–	6	8	11	10	(3)	–	–	–	–	–	–	–	**Total Viscosae**	
									×									quarrei	C.3.i.1
	×		×															ugandensis	„ 2
							×	×	×	×								subulifera var. s.	„ 3α
										×								„ var. polysperma	„ 3β
	×	×	×															subargentea	„ 4
	×	×	×					×	×									microcephala	„ 5
	×	×	×					×										wildiana	„ 6
										×								phyllanthoides	„ 7
	×	×	×			×	×	×	×		×	×	×					vicioides var. v.	„ 8α
	×	×				×	×	×	×				×		×			„ var. rogersii	„ 8β
									×									„ var. occidentalis	„ 8γ
										×								livingstoniana	„ 9
			×															heterocarpa	„ 10
			×															hermannioides	„ 11
			×															taylori	„ 12
–	4	3	7	–	–	2	2	4	5	4	–	–	–	–	–	–	–	**Total Centrae**	

C.3.j. *Atratae*

C.3.k. *Psiloceratiae*

C.3.l. *Geanthae*

	Mauritania*	Senegal and Gambia*	Fr. and Port. Guinea*	Sierra Leone and Liberia*	Ivory Coast*	Fr. Sudan and Upper Volta*	Gold Coast and Br. Togo*	Fr. Togo and Dahomey*	Fr. Niger Colony*	Nigeria and Br. Cameroons*	Fr. Equatorial Afr.	Belgian Congo	A.-E. Sudan and Egypt	Arabia	Asia east of Arabia	Australia	Eritrea	Ethiopia
C.3.j. 1 trachyphylla																		
„ 2 johnstonii												×						
„ 3 fuscosetosa												×						
„ 4α atriceps ssp. a.										×		×						
„ 4β „ ssp. alboglandulosa			×	×						×		×	×					×
„ 4γ „ ssp. setosissima												×						×
„ 4δ „ ssp. glandulosissima																		
„ 4ε ssp. rhodesiaca																		
„ 4θ ssp. kaessneri																		×
„ 5 ramosa												×						
„ 6 schliebenii												×						
„ 7 atricephala																		
„ 8α dasyantha var. d.																		
„ 8β „ var. brevior																		
„ 8γ „ var. viscidior																		
„ 9 glaucifolia												×						
„ 10 setiflora																		
„ 11 melanadenia																		
„ 12 sebungweënsis																		
„ 13 williamsonii																		
„ 14 cecilii																		
„ 15 sanguinea																		
Total *Atratae*	–	–	1	1	–	–	–	–	–	2	–	8	1	–	–	–	–	3
C.3.k. 1 laxeracemosa																		
„ 2 omissa						×		×										
„ 3 longemucronata																		
„ 4 baumiana												×						
„ 5 longiflora																		
„ 6α fulgens var. f.																		
„ 6β „ var. brachybotrys																		
„ 7 sutherlandioides												×						
„ 8α lyallii var. l.																		
„ 8β „ var. nyassica																		
„ 9 gloriosa												×						
„ 10 homblei												×						
„ 11 paucistrigosa																		
Total *Psiloceratiae*	–	–	–	–	–	1	–	1	–	–	–	4	–	–	–	–	–	–
C.3.l. 1 pseudosubulata											×	×	×					
„ 2 thomsonii												×						

Somalia	Uganda*	Kenya*	Tanganyika*	Zanzibar and Pemba*	Madagascar	Port. E. Africa*	Nyasaland*	S. Rhod. and Trop. Bechuanald*	N. Rhodesia*	Angola	S.W. Afr. and N.W. Cape	N. Central Cape and Extra Trop. Bechuanald	Transvaal and Swaziland	Orange Free State and Basutoland	Natal	Eastern Cape	America		
			×			×	×		×									trachyphylla	C.3.j. 1
							×		×									johnstonii	,, 2
							×											fuscosetosa	,, 3
	×	×	×			×	×		×									atriceps ssp. a.	,, 4α
	×	×	×			×	×		×									,, ssp. alboglandulosa	,, 4β
	×	×	×				×		×									,, ssp. setosissima	,, 4γ
			×						×									,, ssp. glandulosissima	,, 4δ
			×						×									,, ssp. rhodesiaca	,, 4ε
	×	×	×				×											,, ssp. kaessneri	,, 4θ
			×															ramosa	,, 5
			×															schliebenii	,, 6
			×															atricephala	,, 7
			×				×											dasyantha var. d.	,, 8α
			×															,, var. brevior	,, 8β
			×															,, var. viscidior	,, 8γ
			×				×		×									glaucifolia	,, 9
						×				×								setiflora	,, 10
								×	×									melanadenia	,, 11
								×	×									sebungweënsis	,, 12
								×										williamsonii	,, 13
						×		×										cecilii	,, 14
						×		×										sanguinea	,, 15
–	4	4	14	–	–	6	9	5	10	1	–	–	–	–	–	–	–	**Total *Atratae***	
			×			×												laxeracemosa	C.3.k. 1
		×	×															omissa	,, 2
									×	×								longemucronata	,, 3
						×												baumiana	,, 4
				×														longiflora	,, 5
			×			×				×								fulgens var. f.	,, 6α
			×						×	×								,, var. brachybotrys	,, 6β
						×												sutherlandioides	,, 7
						×		×										lyallii var. l.	,, 8α
							×											,, var. nyassica	,, 8β
			×															gloriosa	,, 9
	×	×					×											homblei	,, 10
			×															paucistrigosa	,, 11
–	1	2	6	1	–	5	2	1	2	3	–	–	–	–	–	–	–	**Total *Psiloceratiae***	
			×				×		×									pseudosubulata	C.3.l. 1
																		thomsonii	,, 2

C.3.m. *Tinctoriae*

	Mauritania*	Senegal and Gambia*	Fr. and Port. Guinea*	Sierra Leone and Liberia*	Ivory Coast*	Fr. Sudan and Upper Volta*	Gold Coast and Br. Togo*	Fr. Togo and Dahomey*	Fr. Niger Colony*	Nigeria and Br. Cameroons*	Fr. Equatorial Afr.	Belgian Congo	A.-E. Sudan and Egypt	Arabia	Asia east of Arabia	Australia	Eritrea	Ethiopia
C.3.m.1 podocarpa											×	×						
„ 2 ormocarpoides																		
„ 3 garckeana		×										×	×				×	×
„ 4 binderi											×		×					
„ 5α rhynchocarpa var. r.											×	×						
„ 5β „ var. latipinna																		
„ 5γ „ var. uluguruensis																		
„ 6 macrantha																		×
„ 7 roseo-caerulea												×						×
„ 8α emarginella var. e.										×	×	×	×					×
„ 8β „ var. longefoliolata																		
„ 9 emarginelloides																		
„ 10 macrophylla		×	×	×	×		×	×		×								
„ 11α swaziensis var. s.													×					
„ 11β „ var. perplexa																		
„ 12 subcorymbosa																		
„ 13 lupatana																		×
„ 14 spiniflora													×	×			×	
„ 15α subulata var. s.		×		×			×	×		×	×	×	×					×
„ 15β „ var. nubica													×					
„ 15γ „ var. microphylla																		
„ 15δ „ var. scabra										×		×			×			×
„ 15ε „ var. maffeii													×		×		×	×
„ 16 trita														×		×		
„ 17 articulata												×	×				×	
„ 18α coerulea var. c.												×	×	×				
„ 18β „ var. occidentalis									×				×	×	×		×	×
„ 19α bogdanii var. b.																		
„ 19β „ var. peteri																		
„ 20α cryptantha var. c.																		
„ 20β „ var. desmodioides																		
„ 20γ „ var. occidentalis																		
„ 21 amorphoides														×			×	×
„ 22 cavallii																		
„ 23 arrecta		×	×		×	×				×		×	×	×			×	×
„ 24 longeracemosa															×			
„ 25 suffruticosa		×		×	×					×	×	×			×			
„ 26 tinctoria		×	×			×	×	×		×	×		×	×	×			
„ 27 ciferrii																		
„ 28 sparteola																		
„ 29 pungens																		
„ 30 cliffordiana																		
„ 31 dimidiata																		
„ 32 repens												×						
Total *Tinctoriae*	−	6	3	3	2	2	4	3	1	7	(7)	10	13	7	−	−	(7)	(11)

Species (C.3.m.)	Somalia	Uganda*	Kenya*	Tanganyika*	Zanzibar and Pemba*	Madagascar	Port. E. Africa*	Nyasaland*	S. Rhod. and Trop. Bechuanald*	N. Rhodesia*	Angola	S.W. Afr. and N.W. Cape	N. Central Cape and Extra Trop. Bechuanald	Transvaal and Swaziland	Orange Free State and Basutoland	Natal	Eastern Cape	America
podocarpa .. C.3.m.1				×						×	×							
ormocarpoides .. „ 2			×	×		×	×	×	×	×								
garckeana .. „ 3		×	×	×														
binderi .. „ 4			×															
rhynchocarpa var. r. „ 5α				×			×	×	×	×	×							
„ var. latipinna .. „ 5β										×	×							
„ var. uluguruensis „ 5γ				×														
macrantha .. „ 6																		
roseo-caerulea .. „ 7		×	×	×														
emarginella var. e. „ 8α		×	×	×			×	×	×	×	×							
„ var. longefoliolata „ 8β										×	×							
emarginelloides .. „ 9				×						×								
macrophylla .. „ 10																		
swaziensis var. s. .. „ 11α			×	×										×				
„ var. perplexa .. „ 11β			×	×			×		×					×				
subcorymbosa .. „ 12			×	×			×	×	×	×	×							
lupatana .. „ 13			×	×			×	×	×	×				×				
spiniflora .. „ 14																		
subulata var. s. .. „ 15α		×	×	×	×		×	×	×		×	×		×		×		
„ var. nubica .. „ 15β	×																	
„ var. microphylla „ 15γ	×																	
„ var. scabra .. „ 15δ	×	×	×	×	×	×	×	×						×				×
„ var. maffeii .. „ 15ε	×		×	×			×											
trita .. „ 16																		
articulata .. „ 17																		
coerulea var. c. .. „ 18α																		
„ var. occidentalis „ 18β	×		×			×												
bogdanii var. b. .. „ 19α			×	×														
„ var. peteri .. „ 19β				×														
cryptantha var. c... „ 20α								×				×	×	×				
„ var. desmodioides „ 20β					×													
„ var. occidentalis „ 20γ												×						
amorphoides .. „ 21		×																
cavallii .. „ 22	×																	
arrecta .. „ 23		×	×	×		×	×	×	×	×				×		×	×	
longeracemosa .. „ 24		×			×	×												
suffruticosa .. „ 25						×			×		×							×
tinctoria .. „ 26	×		×	×		×	×		×	×								×
ciferrii .. „ 27	×																	
sparteola .. „ 28	×																	
pungens .. „ 29												×						
cliffordiana .. „ 30			×															
dimidiata .. „ 31							×	×	×					×	×	×	×	
repens .. „ 32																		
Total Tinctoriae	(7)	6	15	19	4	–	13	8	13	12	(9)	–	–	–	–	–	–	–

C.3.p. *Alternifoliolae*

C.3.p. *Alternifoliolae*	Mauritania*	Senegal and Gambia*	Fr. and Port. Guinea*	Sierra Leone and Liberia*	Ivory Coast*	Fr. Sudan and Upper Volta*	Gold Coast and Br. Togo*	Fr. Togo and Dahomey*	Fr. Niger Colony*	Nigeria and Br. Cameroons*	Fr. Equatorial Afr.	Belgian Congo	A.-E. Sudan and Egypt	Arabia	Asia east of Arabia	Australia	Eritrea	Ethiopia
C.3.p.1 malongensis												×						
,, 2 intricata														×	×			
,, 3 nephrocarpoides														×				
,, 4 nephrocarpa														×				
,, 5 diphylla	×	×				×		×	×	×	×		×					
,, 6 flavicans																		
,, 7 oblongifolia	×	×						×	×	×			×		×		×	
,, 8α schimperi var. s.													×					×
,, 8β ,, var. baukeana																		
,, 8γ ,, var. crispidula													×					
,, 8δ ,, var. parvifoliolata																		
,, 9 semitrijuga														×			×	
,, 10α daleoides var. dal.																		
,, 10β ,, var. gossweileri																		
,, 11 maritima																		
,, 12 volkensii													×					×
,, 13 insularis													×				×	
,, 14 phillipsiae																		
,, 15 sesquijuga																		
,, 16 viridiflora																		
,, 17 spicata		×	×	×	×	×	×	×		×	×	×	×	×			×	×
,, 18 oxalidea																		
,, 19 kerstingii								×		×								
,, 20 circinella												×						
,, 21 radicifera												×						
,, 22 alternans																		
,, 23 hololeuca																		
,, 24 benguellensis																		
,, 25 auricoma																		
,, 26 antunesiana												×						
,, 27α conjugata var. conjugata			×							×	×	×	×					
,, 27β ,, var. trimorphophylla												×						
,, 27γ ,, var. occidentalis						×				×								
,, 27δ ,, var. schweinfurthii													×					
,, 28 chevalieri											×							
Total *Alternifoliolae*	2	3	2	1	1	3	1	4	2	6	4	7	9	5	–	–	4	3

Species	Ref	Somalia	Uganda*	Kenya*	Tanganyika*	Zanzibar and Pemba*	Madagascar	Port. E. Africa*	Nyasaland*	S. Rhod. and Trop. Bechuanald*	N. Rhodesia*	Angola	S.W. Afr. and N.W. Cape	N. Central Cape and Extra Trop. Bechuanald	Transvaal and Swaziland	Orange Free State and Basutoland	Natal	Eastern Cape	America
malongensis	C.3.p.1																		
intricata	,, 2																		
nephrocarpoides	,, 3																		
nephrocarpa	,, 4																		
diphylla	,, 5																		
flavicans	,, 6							×		×	×	×	×	×	×				
oblongifolia	,, 7											×							
schimperi var. s.	,, 8α		×	×	×			×	×	×	×				×		×		
,, var. baukeana	,, 8β			×	×					×				×					
,, var. crispidula	,, 8γ		×	×															
,, var. parvifoliolata	,, 8δ																		
semitrijuga	,, 9																		
daleoides var. dal.	,, 10α									×	×	×	×	×	×				
,, var. gossweileri	,, 10β											×							
maritima	,, 11											×	×						
volkensii	,, 12		×	×	×														
insularis	,, 13																		
phillipsiae	,, 14	×																	
sesquijuga	,, 15	×																	
viridiflora	,, 16	×																	
spicata	,, 17		×	×	×	×		×	×	×	×	×			×		×		
oxalidea	,, 18									×		×			×				
kerstingii	,, 19																		
circinella	,, 20		×	×	×														
radicifera	,, 21				×						×								
alternans	,, 22											×	×	×	×	×			
hololeuca	,, 23												×						
benguellensis	,, 24											×							
auricoma	,, 25												×						
antunesiana	,, 26				×			×	×	×	×	×							
conjugata var. conjugata	,, 27α		×	×	×							×							
,, var. trimorphophylla	,, 27β										×	×							
,, var. occidentalis	,, 27γ																		
,, var. schweinfurthii	,, 27δ																		
chevalieri	,, 28																		
Total *Alternifoliolae*		(3)	6	7	8	1	–	4	3	7	7	12	6	4	6	1	2	–	–

C.3.n. *Hirsutae*

C.3.o. *Microcarpae*

C.3.q. *Simplices-reflexae*

D. **Indigastrum**

	Mauritania*	Senegal and Gambia*	Fr. and Port. Guinea*	Sierra Leone and Liberia*	Ivory Coast*	Fr. Sudan and Upper Volta*	Gold Coast and Br. Togo*	Fr. Togo and Dahomey*	Fr. Niger Colony*	Nigeria and Br. Cameroons*	Fr. Equatorial Afr.	Belgian Congo	A.-E. Sudan and Egypt	Arabia	Asia east of Arabia	Australia	Eritrea	Ethiopia
C.3.n.1. longebarbata										×		×						×
„ 2α hirsuta var. h.		×	×	×		×	×	×		×	×	×	×		×	×		
„ 2β „ var. pumila										×								
„ 3 astragalina		×				×			×	×		×	×				×	
„ 4 deightonii			×	×						×			×					
Total *Hirsutae*	−	2	2	2	−	2	1	1	1	5	1	3	3	−	−	−	1	1
C.3.o.1 microcarpa		×				×	×		×	×								
C.3.q.1 bongensis											×	×	×					
„ 2 knoblecheri													×					
„ 3 achyranthoides												×	×					
„ 4 hundtii																		
Total *Simplices-reflexae*	−	−	−	−	−	−	−	−	−	−	−	2	3	−	−	−	−	−
D 1 crotalarioides																		
„ 2 fastigiata																		
„ 3 burkeana																		
„ 4α costata ssp. c.		×							×	×			×					
„ 4β „ ssp. theuschii																		
„ 4γ „ ssp. goniodes																	×	×
„ 4δ „ ssp. macra																		
„ 5α parviflora var. p.												×	×		×	×	×	×
„ 5β „ var. occidentalis	×	×																
„ 5γ „ var. crispidula																		
„ 6 argyroides																		
„ 7 argyraea																		
Total **Indigastrum**	1	2	−	−	−	−	−	−	1	1	−	1	2	−	−	−	2	2

Species	Code	Somalia	Uganda*	Kenya*	Tanganyika*	Zanzibar and Pemba*	Madagascar	Port. E. Africa*	Nyasaland*	S. Rhod. and Trop. Bechuanald*	N. Rhodesia*	Angola	S.W. Afr. and N.W. Cape	N. Central Cape and Extra Trop. Bechuanald	Transvaal and Swaziland	Orange Free State and Basutoland	Natal	Eastern Cape	America
longebarbata	.. C.3.n.1		×	×	×			×	×		×	×			×	×	×	×	
hirsuta var. h.	.. ,, 2α		×	×	×	×	×	×	×	×	×	×							
,, var. pumila..	,, 2β											×							
astragalina..	.. ,, 3			×	×			×	×	×	×	×	×		×				
deightonii ..	.. ,, 4																		
Total *Hirsutae*		–	2	3	3	1	–	3	3	2	3	4	1	–	2	1	1	1	–
microcarpa	.. C.3.o.1			×	×		×	×		×	×	×							
bongensis ..	.. C.3.q.1																		
knoblecheri	,, 2																		
achyranthoides	.. ,, 3																		
hundtii ..	.. ,, 4											×							
Total *Simplices-reflexae*		–	–	–	–	–	–	–	–	–	–	1	–	–	–	–	–	–	–
crotalarioides	.. D 1							×											
fastigiata ..	.. ,, 2														×	×	×	×	
burkeana ..	.. ,, 3									×			×	×	×				
costata ssp. c.	.. ,, 4α																		
,, ssp. theuschii	,, 4β											×							
,, ssp. goniodes	,, 4γ		×	×	×														
,, ssp. macra ..	,, 4δ							×		×	×		×		×		×	×	
parviflora var. p. ..	,, 5α		×	×	×					×			×	×	×	×			
,, var. occidentalis	,, 5β										×	×		×					
,, var. crispidula	,, 5γ									×									
argyroides ..	.. ,, 6												×						
argyraea ..	.. ,, 7												×	×		×	×		
Total **Indigastrum**		–	2	2	2	–	–	2	–	4	2	2	5	4	4	3	3	2	–

E **Microcharis**

RHYNCHOTROPIS

	Mauritania*	Senegal and Gambia*	Fr. and Port. Guinea*	Sierra Leone and Liberia*	Ivory Coast*	Fr. Sudan and Upper Volta*	Gold Coast and Br. Togo*	Fr. Togo and Dahomey*	Fr. Niger Colony*	Nigeria and Br. Cameroons*	Fr. Equatorial Afr.	Belgian Congo	A.-E. Sudan and Egypt	Arabia	Asia east of Arabia	Australia	Eritrea	Ethiopia
E 1 spathulata																		
E 2 brevistaminea												×						
E 3 longicalyx			×	×						×	×							
E 4 richardsiae																		
E 5 buchneri																		
E 6 lobata																		
E 7 microcharoides																		
E 8 stipulosa																		×
E 9 hutchinsoniana										×								
E 10α disjuncta var. d.	×								×		×		×	×				
E 10β ,, var. fallax																		
E 11 phyllogramme																		
E 12 pseudoindigofera																		
E 13 annua																		
E 14 contorta														×				
E 15α tritoides var. t.													×	×			×	
E 15β ,, var. obbiadensis																		
E 16 cufodontii																		×
E 17 praetermissa											×	×	×					
E 18α welwitschii var. w.												×	×					
E 18β ,, var. remotiflora										×		×						
E 19 medicaginea																		
E 20 butayei											×	×						
E 21 wajirensis																		
E 22 asparagoides												×						
E 23 tisserantii											×	×						
Total Microcharis	1	–	1	1	–	–	–	–	1	3	5	7	4	3	–	–	1	2
RHYCHOTROPIS																		
1 R. poggei												×						
2 R. marginata												×						
Grand Total	7	39	26	19	12	34	28	19	18	62	(46)	(81)	70	22	–	–	(31)	(50)

Socotra	Somalia	Uganda*	Kenya*	Tanganyika*	Zanzibar and Pemba*	Madagascar	Port. E. Africa*	Nyasaland*	S. Rhod. and Trop. Bechuanaland*	N. Rhodesia*	Angola	S.W. Afr. and N.W. Cape	N. Central Cape and Extra Trop. Bechuanaland	Transvaal and Swaziland	Orange Free State and Basutoland	Natal	Eastern Cape	America	
										×									spathulata .. E 1
																			brevistaminea .. E 2
																			longicalyx .. E 3
				×			×	×		×	×								richardsiae .. E 4
											×								buchneri .. E 5
			×	×	×		×												lobata .. E 6
			×	×															microcharoides.. E 7
																			stipulosa .. E 8
																			hutchinsoniana.. E 9
												×							disjuncta var. d. E 10α
																			,, var. fallax .. E 10β
						×													phyllogramme .. E 11
									×					×					pseudoindigofera E 12
									×										annua .. E 13
																			contorta.. E 14
×																			tritoides var. t... E 15α
	×																		,, var. obbiadensis E 15β
			×	×															cufodontii .. E 16
										×									praetermissa .. E 17
		×		×				×		×	×								welwitschii var. w. E 18α
				×			×	×			×								,, var. remotiflora E 18β
											×								medicaginea .. E 19
		×								×									butayei .. E 20
			×																wajirensis .. E 21
		×	×							×									asparagoides .. E 22
																			tisserantii .. E 23
1	1	3	5	6	1	–	3	3	2	6	5	1	–	1	–	–	–	–	**Total Microcharis**
										×	×								RHYNCHOTROPIS R. poggei .. 1
										×									R. marginata .. 2
0)	(14)	50	72	125	17	–	71	50	72	82	(66)	–	–	–	–	–	–	–	**Grand Total**

INDEX

References in heavy type indicate the position in the present classification where the name is accepted as valid. Those in ordinary type indicate a synonym or a reference in discussion.

Acanthonotus echinatus A1

Anila pentaphylla var. trichocarpa C3d21

Brissonia trapezicarpa B13

Crotalaria marginata Rhynch.2

Cyamopsis psoraloides Cy.1
C. senegalensis **Cy.2**, Cy.3a, Geog.C
C. serrata **Cy.3**, Cy.3a, Geog.C
C. stenophylla Cy.2
C. tetragonoloba Intr.F, **Cy.1**

Dolichos fabaeformis Cy.1
D. psoraloides Cy.1

Eilemanthus strobilifer Cl.11

Galega colutea C3h10
G. esculenta Cy.1

Hedysarum linifolium C3b1
H. nummularifolium A1

Hemispadon pilosus C3g6

Indigastrum deflexum D, D5
Indigastrum macrostachyum D4

Indigofera abyssinica C3m21

I. acanthoclada C3e4
I. accepta C3j10
I. achyranthoides **C3q3**
I. acutifolia E10
I. acutisepala C3d2
I. adami C3d22
I. adenoides C3h5, **C3h8**
I. aeruginis C3b4
I. affinis C3h7, D3
I. alboglandulosa C3j4
I. alta C3m21
I. alternans C3b3, C3p13, **C3p22**, C3p23
I. alternans var. paucijuga C3p13
I. ambelacensis **C3d14**, C3d15
I. amblyantha Geog.B
I. amitina C3d2
I. amorphoides **C3m21**

I. anabaptista B12
I. anceps C3p17
I. andrewsiana **C3b13**
I. angolensis C3m25, E4
I. anil Intr. F, C3m23, C3m25
I. annua E13
I. antennulifera Cl.3
I. antunesiana **C3p26**
I. arabica **B10**, C3b5
I. arenaria B12, C3p22, E10
I. arenaria var. strigosa C3h17
I. arenophila **Cl.3**, Cl.5, Cl.8, Geog.C
I. argentea **C3h17**, C3m17, C3m18, C3p9
I. argyraea **D7**
I. argyroides **D6**
I. arrecta C3j10, C3m20, **C3m23**, C3m26, Econ.1, Econ.5
I. articulata **C3m17**, C3m18
I. aspera Intr.F, B20, **B21**, Geog.C
I. astragalina **C3n3**
I. atrata C3j10
I. atricephala **C3j7**
I. atriceps **C3j4**
I. atriceps ssp. glandulosissima C3j8
I. auricoma **C3p25**
I. australis Econ.3a, Geog.B

I. bagshawei C3b14
I. bainesii C3f4, **C3f5**
I. bakeriana B22
I. bangweolensis **C3h15**
I. baoulensis C2a1
I. barcensis D5
I. barteri **C3h11**
I. basiflora **C3e3**
I. baukeana C3p8
I. baumiana **C3k4**, C3k7
I. baumiana var. hockii C3k7
I. baumiana var. paucijuga C3k7
I. baumiana var. roseo-violacea C3k7
I. benguellensis **C3p24**
I. bequaertii C3h16
I. bergii C3m26
I. berhautiana Cl.9, **Cl.10**, Geog.C
I. biglandulosa **C3g5**
I. binderi C3m3, **C3m4**, C3m10
I. bogdanii **C3m19**
I. boiviniana C3o1

I. filipes C3d2, **C3d3**
I. flaccida C3m15
I. flavicans **C3p6,** Geog.C
I. flavovirens C3m5
I. fragrans C3d21
I. fulgens C3k5, **C3k6,** C3k7
I. fulvopilosa **C3g7**
I. fusca C3n2

I. fuscosetosa **C3j3**

I. gairdnerae **Cl.2**
I. galegoides Econ.3a
I. galpinii C3d2
I. garckeana **C3m3,** C3m5
I. geminata C3d, C3d22, **C3d23**
I. gerrardiana Intr. F,
I. gilletii C3d1
I. glandulosa Econ.6
I. glabra C3d21, C3g
I. glauca C3m17
I. glaucifolia C3j4, **C3j9**
I. gloriosa **C3k9**
I. glutinosa C3h10, C3h16
I. goetzei C3d26
I. goniocarpa C3m13
I. goniodes D4
I. goniodes var. damarensis D3
I. grata C3h7
I. grata var. longeracemosa C3i6
I. graveolens C3n2
I. griquana C3n1
I. grisea C2a18, C3h18
I. griseoides **C3b11**
I. guatimalensis Econ.1
I. guineensis C3g6
I. gyrocarpa C3m5

I. hedranophylla B3
I. hedyantha **C3d26**
I. hendecaphylla C3m18, C3p13, C3p17,
    Econ.3c
I. heptaphylla D4
I. hermannioides **C3i11,** C3i12
I. heterocarpa C3b, C3h1, **C3i10**
I. heterotricha **C3h4,** C3h5
I. heudelotii **C3d4,** C3k
I. heudelottii var. elliotii C3d5
I. hewittii C3h10, **C3h14**
I. hilaris C3b, **C3h1,** C3i, C3i10, Econ.3b
I. hirsuta C3n, C3n1, **C3n2,** C3n3, Econ.3b,
    Econ.4b, Econ.5
I. hirsuta var. polystachya C3n1
I. hislopii C3i8
I. hochstetteri B5, **B12,** B13, B15
I. hockii C3h1
I. hofmanniana C3m25
I. hololeuca **C3p23**
I. hololeuca var. angolensis C3p10
I. holstii C3j4
I. holubii **B7**
I. homblei **C3k10**
I. houer C3m18, C3m26

I. hundtii **C3q4**
I. hutchinsoniana E, **E9**
I. hybrida C3h1

I. inhambanensis **C1.8**
I. insularis **C3p13**
I. intricata C3e, C3e2, C3f6, **C3p2**
I. inyangana C3d26
I. ischnoclada **Cl.1,** Cl.2

I. jamaicensis C3m15
I. jaubertiana B12
I. johnstonii **C3j2**
I. junodii C3h10

I. kaessneri C3j4
I. kandoensis C3i4
I. karongensis C3i4
I. kelleri B17
I. kengeleensis C3d1
I. kerensis C3f3
I. kerstingii **C3p19**
I. kirilowii Intr.F, C3k, Geog.B.
I. kirkii **C3g2**
I. kisantuensis C3m23
I. kleinii C3p17
I. knoblecheri **C3q2,** C3p27
I. komiensis C3b9
I. kongwaensis **C3d15**
I. kuntzei **Cl.4**

I. lanuginosa Cl.11
I. lasiantha **B18,** B19, C3m
I. lateritia C3n2, E3
I. latibracteata C3d8
I. latifolia E6
I. latipinna C3m5
I. latisepala C2a15, **C2b3**
I. laxeracemosa C3k, **C3k1**
I. leendertziae **B6**
I. lepida C3d2
I. leprieurii **C3b8**
I. leptocarpa E10, E15
I. leptoclada **C2a16,** C2a17
I. letestui **B1**
I. linearis D5
I. linearis var. sessilis B20, B21
I. linifolia Intr.E, **C3b1,** Econ.3b, Econ.6
I. livingstoniana **C3i9**
I. lobata **E6**
I. lonchocarpifolia C3m5
I. longebarbata **C3n1,** C3n2
I. longemucronata **C3k3**
I. longeracemosa C3i6, **C3m24**
I. longicalyx **E3**
I. longiflora **C3k5**
I. longipes C3d2
I. longispina **C3e4,** Geog.C
I. lotoides C3p7
I. lotononoides C2a4, **C2a10**
I. lupatana **C3m13**
I. lupulina Cl.11

I. lyalii **C3k8**
I. lydenbergensis **C3h5,** C3i

I. macra C3h12, D4
I. macrantha C3k, **C3m6,** C3m7
I. macrocalyx **C2a4,** C2a5
I. macrocarpa C3b8
I. macrophylla C3k, **C3m10,** C3m11
I. macrostachya D4
I. maffeii C3m15
I. malongensis **C3p1**
I. manyoniensis **C3d9**
I. marginulata C3m15
I. maritima **C3p11**
I. masonae C3d2
I. masukuensis C3j4
I. mearnsii C3k, C3m11
I. medicaginea D, **E19,** E20
I. megacephala **C2a6**
I. melanadenia C3j, **C3j11,** C3j12
I. melolobioides B3
I. mendonçae **C3d17**
I. micrantha C3m25
I. microcalyx **C3b14**
I. microcarpa C3b3, C3o, **C3o1**
I. microcephala C3h3, **C3i5**
I. microcharoides Intr.B, **E7,** E16, E21
I. micropetala **C3d10**
I. microscypha C3b15
I. mildbraediana **C3b12,** C3b13, C3i3
I. milne-redheadii **C3h3**
I. mimosoides **C3h7,** C3i3
I. minimifolia C3d21
I. mittuensis B1
I. moeroensis C3h1
I. mollicoma C3j14
I. monantha C2b, **C3d7**
I. mossambicensis B23
I. mounyinensis C3i3
I. mucronata C3m15
I. multifoliolata C3h10
I. multijuga C3d1
I. mwanzae **C3d12**
I. mysorensis C2

I. nairobiensis C3d11, **C3d25,** C3d25A
I. nambalensis C3m5
I. nebrowniana **C3f4,** C3f5
I. neglecta C3p17
I. nematopoda C3h12
I. nephrocarpa C3p3, **C3p4**
I. nephrocarpoides **C3p3**
I. nigricans **C2a15,** C2b3
I. nigritana **C2b2**
I. noldeae C3k7
I. nummularia **C3g4**
I. nummulariifolia **A1**
I. nyikensis C3h1

I. obermejerae C3k8
I. oblongifolia **C3p7,** C3p8, Econ.4a, Geog.B
I. ogadensis **C3f2**

I. oligantha C2b2
I. oligophylla **B9**
I. oligosperma C3h16
I. oliveri C3m11
I. omissa C3k, **C3k2**
I. onobrychoides C3p17
I. ormocarpoides **C3m2**
I. ornithopodioides B12, C3m26
I. orthocarpa C3m26
I. oubanguiensis **C2a16**
I. oxalidea **C3p18**
I. oxytropis C3j14

I. palustris C3d1
I. paniculata **C2a3**
I. paracapitata **C2a8**
I. parkeri C3p17
I. parviflora C2a16, D, **D5**
I. parvula C3p13, C3p17
I. patens B3, Econ.3c
I. patula C3h1
I. paucifolia C3p7, C3p13, Econ.4a
I. pauciflora C3d14
I. paucistrigosa **C3k11**
I. pearsonii C3h5
I. pechuelii C3h5
I. pectinata C3p17
I. pentaphylla C3d6, C3d21
I. perplexa C3m11
I. perrottetii C301
I. petiolata **C3g1**
I. phillipsiae **C3p14**
I. phyllanthoides C3d1, **C3i7**
I. phyllogramme **E11**
I. pilosa C3d21, C3g, C3g5, **C3g6,** C3g7,
    Econ.4a, Geog.C
I. pityophila C3m20
I. pobeguinii **C3b15**
I. podocarpa **C3m1,** C3m10
I. podophylla **C3a1**
I. poggei Rhynch.1
I. polycarpa Cl.3, Cl.8
I. polysperma C3b7
I. polysphaera **C2a1**
I. praetermissa D, **E17,** E18
I. praticola **B14,** B15, Geog.C
I. preladoi C4g2
I. pretoriana C3j10
I. prieureana **C3b9**
I. procera C2a3, C2a17, C3b6
I. psammotropha Cy.3
I. pseudo-indigofera **E12**
I. pseudointricata **C3f6,** C3f7, C3p2
I. pseudosubulata **C3LI**
I. pseudotinctoria Intr.F.
I. psilostachya B18
I. pulchella C3k8
I. pulchra **C2a13,** C2a14
I. pungens C3e, **C3m29**
I. pusilla C3p17

I. quarrei **C3i1**

I. quartiniana C3m15

I. radicifera C3p18, **C3p21**
I. ramosa **C3j5**
I. rautanenii **B4**
I. rehmannii C3j14
I. relaxata B20
I. remotiflora E18
I. repens **C3m32**
I. retroflexa C3m15
I. retusa C3m17
I. rhynchocarpa C3m, **C3m5**
I. rhynchocarpa var. quadrangularis C3m3
I. rhytidocarpa **B5**
I. richardiana C3o1
I. richardsiae **E4**
I. rogersii C3i8
I. roseo-caerulea **C3m7**
I. rostrata D2
I. rothii **C3e1**
I. rudis C3h4
I. rufescens C2a13
I. ruspolii **C3f1**, C3f2, C3p
I. rutschuruensis C3m15

I. sanguinea C3j14, **C3j15**
I. saxicola D6
I. scaberrima B20
I. scabra C3m15
I. scarciesii **C3c1**
I. schimperi **C3p8**, Econ.4a
I. schinzii Cd2
I. schlechteri C3n1
I. schliebenii **C3j6**
I. schweinfurthii C3p27
I. scopa C3m23
I. sebungweënsis **C3j12**
I. secundiflora **C3h16**
I. secundiflora var. glandulosissima C3j4
I. semhaensis B12 E10
I. semitrijuga C3h17, **C3p9**
I. semitrijuga var. tetrasperma C3h17
I. semlikiensis C3i8
I. senegalensis **B13**, B14, B18, C3b6, Geog.C
I. sericea Cl.3, Cl.10
I. sericea forma australis Cl.6
I. sesbaniifolia C3d1
I. sesquijuga **C3p15**
I. sesquijuga var. obbiadensis E15
I. sessiliflora Intr.F, **C3b5**, C3p
I. sessilifolia B, **B3**, C3e, Econ.3c
I. seticulosa C3h10
I. setiflora **C3j10**, C3j12
I. setosissima C3j4
I. shinyangensis C3i5, C3j4
I. shirensis C3h7
I. similis C3b15
I. simplicifolia C3b, C3b6, **C3b7**, C3b8,
C3f5
I. sisalis **B15**
I. smithioides C3g2
I. sofa C3d4
I. somalensis C3p9

I. sordida **C3h2**, C3h5, C3p
I. sousae C3j4
I. spachii C3m15
I. sparsa C3d6, **C3d18**, C3d22
I. sparsiflora C4f3
I. sparteola C3m14, **C3m28**
I. spathulata E, **E1**
I. spicata C3m18, C3p, C3p13, **C3p17**,
Econ.3c, Econ.4a, Econ.5
I. spiniflora C3e, C3e2, **C3m14**, C3p2, D1
I. spinosa **C3e2**, C3e3, C3e4, C3m14, C3p2,
Geog.C
I. spinosa var. spiniflora C3m14, E15
I. spirocarpa C3p20
I. stenophylla **C3b10**, C3b13
I. stenophylla var. brachypoda C3b9
I. stenophylla var. latifolia C3b9
I. stenophylla var. nyassae C3p26
I. stipulosa **E8**
I. strigulosa C3m15
I. strobilifera Cl.5, Cl.10, **Cl.11**
I. stuhlmannii B15
I. suaveolens C3d21, C3f, **C3f3**, C3f4
I. suaveolens var. pseudocongolensis C3d21
I. suaveolens var. subquadriflora C3d21,
C4f3
I. subargentea C3b, C3h3, C3h10, **C3i4**,
C.3i5
I. subcorymbosa C3m12
I. subhirtella C3p12
I. subincana C3m15
I. subquadriflora C3d21, C3f3
I. subulata C3m14, **C3m15**, Econ.4a
I. subulata var. scabra C3m31
I. subulata var. subulata C3m15
I. subulifera C3b, C3b12, **C3i3**
I. suffruticosa Intr.F, C3m18, **C3m25**,
Econ.1, Econ.3b, Econ.5
I. sumatrana Intr.F, C3m26, Econ.5
I. supralevis C3p18
I. sutherlandioides C3b9, C3k, C3k4, C3k6,
C3k7
I. swaziensis **C3m11**

I. taborensis **Cl.9**
I. tanaensis **C3f7**
I. tanganyikensis C3d9, **C3d11**, C3d12,
C3d13, C3d17, C3d25A
I. tanganyikensis X nairobiensis **C3d25A**
I. taylori **C3i12**
I. tenella B21, E, E9
I. tenuicaulis C3i8
I. tenuis **C3d19**
I. tenuisiliqua E15
I. terminalis **C2a5**
I. tetragona C3m3
I. tetragonoloba C3m15
I. tetrasperma **C3b6**, C3b7
I. tetrasperma var. hexasperma C3b7
I. tettensis C3p8, Econ.4a
I. theuschii D4
I. thomsonii C3k, C3L, **C3L2**

I. thonningii C3m15
I. tinctoria Intr.F, C3m, C3m18, C3m23,
    C3m26, Econ.1, Econ.3b, Econ.5
I. tisserantii D, **E23**
I. torrei **C3g8**
I. torulosa B6, **B8,** C3m2
I. torulosa var. angustiloba B9
I. trachyphylla C2a1, **C3j1,** C3j2
I. transvaalensis C3i8
I. trialata C2, **C2a2**
I. tribuloides C3b5
I. trichopoda C2b, **C2b1**
I. trigonelloides **C3b4,** C3b5, C3p, Geog.C
I. trimorphophylla C3p26, C3p27
I. trita C3e, C3m14. C3m15, **C3m16,** D1
I. tritoides C3m14, E14, **E15,** E16, E21

I. ugandensis C3i2
I. uhehensis C3d21, **C3i2**
I. umbonata C3m23
I. umbraticola C3m15
I. uncinata C3m25
I. uniflora C2b

I. vanderystii **C2a14**
I. variabilis C3f4, C3f5, E18
I. velutina C3j14
I. vicioides **C3i8,** C3k, C3k1
I. vicioides var. rogersii C3j10
I. viscidissima **C3h9**
I. viscosa C3h, C3h7, C3h10, C3n2
I. viscosa var. brachycarpa C3h10
I. viscosa var. dembianensis C3h10
I. viscosa var. subglabra C3d21
I. viridiflora **C3p16**
I. vohemarensis C3d11, **C3d21,** C3f3

I. volkensii **C3p12**

I. wajirensis **E21**
I. wauensis C3q3
I. welwitschii D, E2, E9, E17, **E18**
I. welwitschii var. remotiflora E2
I. welwitschii var. simplicifolia E10
I. wentzeliana C3h1
I. wildemani C3d14
I. wildiana **C3i6**
I. williamsonii **C3j13**
I. wilmaniae **B16**
I. wituensis **C3d16,** C3d17

I. zavattarii **C3d24**
I. zenkeri C3h10
I. zigzag C3h1

Lupinus trifoliolatus Cy.1

Microcharis angolensis E4, E5
M. galpinii E12
M. latifolia E6
M. pseudo-indigofera E12
M. tenella E3, E9

Psoralea tetragonoloba Cy.1

Rhynchotropis curtisiae Intr.B, E7
R. dekindtii Rhynch.1
R. marginata **Rhynch.2**
R. poggei **Rhynch.1**
R. praecox Rhynch.2

Sphaeridiophorum abyssinicum C3b1
S. linifolium C3b1

Tephrosia crotalarioides C3m14, D1

Printed in Great Britain under the authority of HER MAJESTY'S STATIONERY OFFICE
by The Campfield Press, St. Albans